ENVIRONMENT AND THE EXPERIMENTAL CONTROL OF PLANT GROWTH

EXPERIMENTAL BOTANY
An International Series of Monographs

CONSULTING EDITORS

J. F. Sutcliffe

School of Biological Sciences, University of Sussex, England

AND

P. Mahlburg

Department of Botany, Indiana University, Bloomington, Indiana, U.S.A.

ENVIRONMENT AND THE EXPERIMENTAL CONTROL OF PLANT GROWTH

R. J. DOWNS

North Carolina State University,
North Carolina, U.S.A.

and

H. HELLMERS

Duke University, Durham,
North Carolina, U.S.A.

1975

ACADEMIC PRESS · LONDON · NEW YORK · SAN FRANCISCO
A Subsidiary of Harcourt Brace Jovanovich, Publishers

ACADEMIC PRESS INC. (LONDON) LTD.
24/28 Oval Road,
London NW1

United States Edition published by
ACADEMIC PRESS INC.
111 Fifth Avenue
New York, New York 10003

Library of Congress Catalog Card Number: 74 18524
ISBN: 0 12 221450 1

PRINTED IN GREAT BRITAIN BY
J. W. ARROWSMITH LTD., BRISTOL

Preface

The biologist now is becoming aware that environment plays such an important role in plant metabolism, growth and development, that research in these areas needs to be done in a known environment which in the final analysis is controlled environment. Therefore, during the past 25 years there has been a large increase in interest and use of controlled environments in biological research. The demand for equipment that can provide the desired conditions has brought about a flurry of plant growth chamber building by manufacturers, some as newly formed branches of large manufacturing companies and others as small independent companies, and by individuals making their own units. This has resulted in a large number of units that vary in size, reliability and particularly in the degree of control for the various parameters of the factors that affect plant growth. On one hand the conditions requested by some biologists cannot be met without exorbitant cost and on the other hand some accept whatever the manufacturer provides and complain later. Unfortunately, the researcher is often concerned with only one or two factors of the environment and therefore, he often fails to consider in his chamber specifications the other factors that can have profound effects on the research results.

The purpose of this book is to bring together information about the environmental factors in terms of their general role in plant growth and methods of providing the desired levels and limits of control. It is hoped that this information will help the biologist to understand what he is buying or constructing in terms of environment variability in plant growth facilities. Also, we hope, the book will provide some help and guidance to those who have encountered the problem of not obtaining the degree of control they had expected in the units they have in hand. It must be remembered that in a system as complex as the environment there has to be some compromise in the degree of control and the extent of the compromise is often dictated by cost. Therefore, it is important to know what these compromises are and the degree of control that can be obtained for each factor.

Because we are credited with writing the book, we must also assume the responsibilities for any omissions, errors or other faults. Nevertheless, we do thank all those who contributed to the volume through their research suggestions or critical review. The book would never have been finished without the enduring patience and noble efforts of Mrs Fleming and Mrs Wilkes in typing, proofreading and general handling of the manuscript.

Last but not least our thanks go to our wives whose encouragement and understanding always added that something extra that is needed in a project such as this.

April, 1975   R.J.D.
                                                                  H.H.

Contents

CHAPTER I

Introduction

Environment control for plant growth is not a particularly new idea. In fact, attempts to control the environment of the plant began with the development of agriculture. The land is cleared of all competing species and replanted with the single type desired by the grower. Competition from undesirable plants is prevented by a vigorous program to destroy them. Irrigation, a practice as old as agriculture, is aimed at changing the natural water relations of an area just as the addition of fertilizer alters the nutritional levels of the soil.

Centuries ago people began bringing exotic plants indoors to avoid cold weather damage. This practice led quite naturally to the development of the greenhouse. Off-season production of speciality crops in artifically warmed greenhouses developed over the years along with the use of such facilities in research programs. Year-round use of the greenhouses created a demand for cooling as well as heating. Photoperiod control and efforts at supplementing the low light intensities of winter are commonplace. Some greenhouse systems modify the carbon dioxide level making the concentration higher than normal.

As biological research became more sophisticated, solutions to important problems were unattainable because the naturally fluctuating environment obscured experimental results. In physics and chemistry precise control of environment has long been recognized as essential to experimentation and more recently the biologist has become aware of the importance of the need for controlled environments in which to conduct his research. However, the biologist has been repeatedly confronted with his own lack of knowledge of how the various environmental factors affect the biological organism, a situation that is further complicated by genetic variation even within species.

Air-conditioned greenhouses, plant growth chambers and phytotrons are the natural result of a progression of events leading step by step to precise environment control. The large breakthrough came with the designing of the first phytotron in 1949 by F. W. Went (Went, 1957) at the California Institute of Technology. Since then much progress has been made in improving the efficiency of such systems and the reliability of the equipment. Each of the many phytotrons that has been built since 1949 has incorporated new ideas and modifications for specific types of research (Braak and Smeets, 1956; Alberda, 1958; Morse and Evans, 1962; Zscheile *et al.*, 1965; Wettstein, 1967; Kramer *et al.*, 1970; Rajki, 1971; Downs *et al.*, 1972; Nitsch, 1972; Konishi, 1972).

Many advances also have been made in the construction and operation of individual chambers. Nevertheless the biologist that buys one or more plant growth chambers is often dismayed at the mechanical and electrical problems encountered.

Controlled Environments as Research Tools

Plants are highly buffered systems as regards their response to the environment. Thus, the optimum for any given factor may, and usually does, shift with changes in the other environmental factors. Therefore, it is almost mandatory that the factors, other than the one being tested, be held constant over the range of the test factor. If the other factors cannot be held constant, their degree of variability must be taken into consideration in evaluating the results.

Phytotrons are facilities consisting of multiples of similar plant growth chambers and controlled greenhouses. For instance studies using up to thirty combinations of temperature have been used in several phytotrons to determine near optimum conditions for the growth of plants (Hellmers *et al.*, 1970; Hellmers and Rook, 1973; Greenwald, 1972). A very extensive interaction study of temperature and photoperiod with tomato plants was conducted by Kristoffersen (1963). The uses of controlled environment can be grouped into two major types. One use involves growing plants under reproducible conditions for growth, development and biochemical studies. The second use involves growth and development of plants simultaneously over a range of an individual environmental factor as well as studying the interaction of several factors.

As soon as the biologist begins to use controlled-environment facilities, decisions must be made for which he may have few data. Whereas with less elaborate facilities he could decide on a minimum temperature and perhaps have some selection of photoperiod, other factors like light intensity, maximum temperatures, humidity, atmospheric composition and air flow had to be accepted as they occurred. If temperature is to be studied in a controlled facility, for example, over what range and in what steps should the experiment be designed? Should the night temperature be different from that during the day and if so how much different? Many other environmental factors may also need decisions before a study can begin. These include: How much light energy should be used? What photoperiod should be selected? Is humidity control important? Does the CO_2 level need to be maintained? After the aerial environment is selected then there are problems of the root environment. Also water relations and nutrition levels become critical because the plants may grow at a quite different, usually faster, rate than that which the investigator ordinarily obtains.

The advantages and the problems encountered in the use of plant growth facilities seem to be almost limitless. Whether the advantages outweigh the problems or vice versa depends upon the time when the question is raised. Whether or not the experiment is progressing as anticipated usually dictates the answer. Frequently, the problems encountered could be anticipated or detected early and serious effects avoided. The objective of the present book is to outline some of the advantages, complexities and problems encountered in growing plants in controlled environments and to offer what we believe to be helpful suggestions in the operation and use of such facilities.

Construction and Operation of Controlledn-Evironment Facilities

The basic principles of physics involved in controlling the environment in plant growth facilities are described in several publications. The book entitled "The Experimental Control of Plant Growth" by F. W. Went (1957) explains in great detail the construction and physical operation of the first phytotron at the California Institute of Technology. The two-unit phytotron on the Duke University and the North Carolina State University campus has been the subject of three papers (Hellmers and Downs, 1967; Kramer et al., 1970; Downs et al., 1972). The first outlines the criteria developed for the various parameters of the environment in the planning stage and how it was anticipated the specified degree of control would be achieved. The second paper describes the two units in detail and was written after their construction. The third paper, a critical review of the mechanical systems both in terms of what would and would not be recommended for future use, was written after several years of operation experience.

The perfect plant growth chamber has yet to be built. As one plant growth chamber manufacturer stated: "We thought we built the perfect chamber when we made the first one and have had the same thought about each chamber since then but we have never built two alike. We continually find ways to improve them." Each facility, be it the component of a phytotron or a single plant growth chamber, has its own characteristics and idiosyncrasies. Nevertheless, some problems in both the mechanical operation of the unit and the growth of plants in the unit arise repeatedly. While few people report their problems some evidence of them is to be found in the literature. Over the years through the use of controlled environments, and in conversation with others who use them, we have accumulated information on many of the problems. Therefore, this book includes some suggestions and guidance that we hope will make it possible for researchers to become more aware of the problems and how to avoid them.

This book is oriented toward the advantages and problems encountered in the use of controlled environments more than have previous books on the subject. Nevertheless, the information contained should be considered as a supplement to up-date the information reported in such works as R. O. Whyte's *Crop Production and Environment* (1960), F. W. Went's *Experimental Control of Plant Growth* (1957) and L. T. Evans' *Environmental Control of Plant Growth* (1963).

References

Alberda, T. (1958). The phytotron of the Institute for Biological and Chemical Research on field crops and herbage at Wageningen. *Acta Botanica Neerlandica* **1,** 265–277.

Braak, J. P. and Smeets, L. (1956). The phytotron of the Institute of Horticultural Plant Breeding at Wageningen, the Netherlands. *Euphytica* **5,** 205–221.

Downs, R. J., Hellmers, H. and Kramer, P. J. (1972). Engineering problems in the design and operation of phytotrons. *American Society of Heating, Refrigerating and Air-Conditioning Engineers Journal* **14,** 47–55.

Evans, L. T. (1963). "Environmental Control of Plant Growth", Academic Press, London and New York.

Greenwald, S. M., Sr. (1972). "Some environmental effects on the growth and monoterpene production of *Pinus taeda* L. and *Ocimum basilicum* L." Ph.D. Thesis, Duke University, Durham, North Carolina.

Hellmers, H. and Downs, R. J. (1967). Controlled environments for plant-life research. *American Society of Heating, Refrigerating and Air-Conditioning Engineers Journal* **9,** 37–42.

Hellmers, H. and Rook, D. A. (1973). Air temperature and growth of radiata pine seedlings. *New Zealand Journal of Forestry Science* **31,** 271–285.

Hellmers, H., Genthe, M. K., Ronco, F. (1970). Temperature effects on growth and development of Engelmann Spruce. *Forest Science* **16,** 447–452.

Konishi, M. (1972). Phytotrons in Japan and the Japanese Society of Environment Control in Biology and its activities including the plan of the National Biotron Center. *Environmental Control in Biology* **10,** 1–10.

Kramer, P. J., Hellmers, H. and Downs, R. J. (1970). SEPEL: new phytotrons for environmental research. *BioScience* **20,** 1201–1204.

Kristoffersen, T. (1963). Interactions of photoperiod and temperature in growth and development of young tomato plants (*Lycopersicon esculentum* Mill.) *Physiologia Plantarum Supplementum* **1.**

Morse, R. H. and Evans, L. T. (1962). Design and development of CERES—an Australian phytotron. *Journal of Agricultural Engineering Research* **7,** 128–140.

Nitsch, J. P. (1972). Phytotrons: past achievements and future needs. *In* "Crop Processes in Controlled Environments". (A. R. Rees, K. E. Cockshull, D. W. Hand and R. G. Hurd, eds), pp. 33–55. Academic Press, London and New York.

Rajki, S. (ed.) (1971). "The First Twenty Years of Martonvasar." Agricultural Research Institute of the Hungarian Academy of Sciences, Martonvasar.

Went, F. W. (1957). "The Experimental Control of Plant Growth." Chronica Botanica Co., Waltham, Mass.

Wettstein, D. von. (1967). The phytotron in Stockholm. *Studia Forestalia Suecica*, Royal College of Forestry, Stockholm, Sweden. Nr. **44,** 1–23.

Whyte, R. O. (1960). "Crop Production and Environment." Faber and Faber, London.

Zscheile, F. P., Anderson, S. M., Leonard, A. S., Neubauer, L. W., and Sluka, S. J. (1965). A sunlight phytotron unit as a practical research tool. *Hilgardia* **36,** 493–565.

CHAPTER II

Temperature

Definition and Measurement

Temperature is one of the key factors controlling the growth of plants. Temperature operates directly through the control of biochemical reaction rates in the various metabolic processes and indirectly through the development of water stress caused by the physical process of transpiration.

Temperature is defined in terms of the absolute temperature scale established by Kelvin in 1848. A hundred years earlier, 1742, Celsius proposed the temperature scale most universally used in biology today.

The Celsius or Centigrade scale, the Kelvin scale and the constant volume, hydrogen gas thermometer scale used by the U.S. Bureau of Standards all define the magnitude of 1 degree as $\frac{1}{100}$ of the difference between the temperature of melting ice and that of boiling water under one atmosphere or 760 mm of mercury pressure. The Kelvin scale is based on absolute zero. Consequently $0°C$ equals $273·16$ K and $100°C$ equals $373·16$ K.

Another scale for temperature was established by Fahrenheit about 1724 and is based on a zero point equal to the temperature of a mixture of ice and salt and $100°F$ set at supposedly body temperature. However, the latter obviously was in error as it is too high for normal body temperature: the person used as the standard must have had a fever. Water freezes at 32°F and boils at 212°F. Thus a degree F is $\frac{5}{9}$ the magnitude of a degree C. The Fahrenheit scale is rarely used in science and is mentioned here only because it has persisted in the engineering and meteorological fields.

SELECTING A METHOD OF MEASUREMENT

Selecting a method of determining temperature depends largely upon the sensitivity and response time desired. The temperature range and where it is to be measured would also influence the choice. In controlled environment applications using biological material the range is relatively small, rarely exceeding −10 to 55°C. The site of the temperature to be measured may be in or on the biological sample, at some depth in a substrate or simply air temperature.

Temperature detecting systems operate through a measurable change in the condition of the sensor. Temperature-dependent outputs of the system may include:

1. change in volume as in a liquid–glass thermometer;
2. displacement caused by differential expansion of the metals in a bimetal strip;
3. displacement caused by the volume change of a gas in a bellows or tube;
4. change in electric current generation;
5. change in electrical resistance.

There are other outputs, of course, like the paramagnetic changes in salts of iron ammonium chloride used primarily below 4 K or the degree of ionization of a gas used for very high temperature measurements. However, for biological studies the ones listed are the most adaptable to controlled-environment conditions.

Any measuring system will of course influence the process to be measured and this interference must be kept small. For instance, the sensor output and the readout device may interact, or the system may be sensitive to the influence of factors of the environment other than the one to be measured.

In controlled-environment applications one might argue that the larger the sensor the better the indication of the average temperature of the space. One might also argue that sensitivity does not need to be high and that a slow response time is more imitative of plant response. Perhaps these viewpoints are valid ones but not if the purpose of the measurement is to describe the physical nature of the environment. By sensitivity we mean the transfer ratio: output quantity/input quantity or $dQ/dT = S$. Response

speed or time constant of the system refers to time lag between a change
in conditions and the response of the instrumentation. For example, a
common laboratory thermometer or a thermograph, if calibrated properly,
can accurately detect a temperature change of about $\frac{1}{2}$°C providing the
change occurs over a relatively long time span. If temperature changes
occur rapidly as in most controlled-environment chambers a 1°C fluactu-
tion will record as less than $\frac{1}{4}$°C because of the slow response. A thermocouple, thermistor or resistance thermometer under the same conditions
record the change as nearly 1 degree; providing of course that the recorder
response is sufficiently faster than the sensor so that the system time
constant equals that of the sensor.

In controlled-environment rooms we should be concerned about
measuring the air temperature because it adequately describes one phase
of the environment, and if made correctly should allow comparisons
between different experiments. Radiant temperature, which will be
discussed later, should be a separate measurement. The sensor selected
to measure air temperature should be small because the greater the mass
or surface area the more difficult it is to shield from other factors, mainly
the high intensity light, that can influence the measurement. Shielding the
sensor in an insulated or highly reflective housing through which room air
is drawn rapidly works very well. In our phytotron the aspirated sensor
housings are on a long lead so they can be placed anywhere in the room.
Our experience has been that the aspirated system provides an average of
the room air temperatures so we find little difference due to position.

The best methods for temperature measurement in plant growth controlled-environment rooms are thermocouples or thermistors. Thermo-
graphs have too slow a response and take up too much space, especially if
they are properly shielded. Gas- or liquid-filled pneumatic type sensors
are objectionable for much the same reasons. All the mechanical type
instruments must finally be calibrated, usually against a thermocouple or
thermistor.

Resistance thermometers may be theoretically ideal but in controlled-
environment room practice they have proved troublesome. Systems that
might be more reliable are also expensive. Thermocouples, however, can be
made by the investigator and the many types of excellent readout and
recording devices make such a system extremely flexible. Iron–constantan
and copper–constantan thermocouples are the most commonly used, but
under some conditions the iron has a tendency to corrode. The high
thermal conductivity of copper may be undesirable under certain cir-
cumstances. Chromel–alumel would then be a better choice. At NCSU,
however, we have used copper–constantan without difficulty and at
Duke iron–constantan has not produced any real problems due to
corrosion. Thermistors are now available with nearly linear characteristics

and excellent stability. They can be used quite satisfactorily for environmental measurements.

Thermocouples are our choice, however, because they can be used to measure the temperature of soil, water and plant tissue as well as air. Moreover, when we attempt to measure radiant temperature the thermocouple is a very adaptable tool. Radiant temperature is the total temperature potential of a given environment including heat radiated from walls and the radiant heat from the light source. Estimates of this value might be made using the difference in temperature between highly absorptive and highly reflective bodies. Read *et al.* (1963) reported on the use of steel balls and thermocouples for this purpose, while Funk (1963) reported on net radiometers using ribbon thermopiles that show the difference between long and short wavelength radiation.

Read's method measures the total thermal environment and does not measure radiant temperatures *per se*. The Sutton-McNall (1964) two sphere radiometer, however, electrically heats the absorbing and reflecting spheres to eliminate convection differences. The difference in electrical energy required to maintain the two spheres at the same temperature is then proportional to the radiant temperature.

Heat Balance of Plants in Controlled Environments

A plant can be considered as almost a black body in terms of heat exchange balance. Energy is absorbed across the spectrum except for a small nonabsorbing window in the green band and another in the infrared area (Fig. 1). At a certain temperature the heat transfer into and out of the plant balances to zero. If this temperature is above or below the maximum and minimum cardinal points for the plant it is killed. The amount of light energy used in photosynthesis is usually less than 5 %, and the heat produced by the various metabolic processes is usually so small that these do not need to be considered in the energy balance. Due to the various heat loss processes leaf temperature may vary more than 10°C above or below that of the ambient air.

Energy, either as heat or as light, much of which is converted to heat, is received by the plant from many sources.

The amount of heat energy received from the light source will vary with the type of lamps being used. For instance, incandescent lamps emit more energy in the infrared wavelengths than do fluorescent lamps. Xenon lamps also produce a lot of radiation in the heat bands as well as in the visible spectrum. Most of the radiant energy incident upon plants is reflected, transmitted or fluoresced. The remainder is absorbed and causes a rise in plant temperature and an increase in the rate of heat loss either through reradiation or transpiration. The rise in leaf temperature due to radiant energy from a light source is evident in Fig. 2.

The difference in temperature between the walls of the chamber or greenhouse and the plant determines the balance of radiant energy flow between them. If the walls are warmer they will add to the radiant energy received by the plant and thus cause a heating of the plant, yet almost no attention is paid to this kind of energy balance in evaluating controlled-environment facilities.

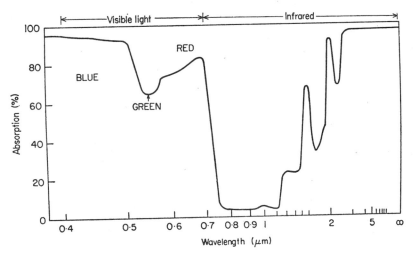

FIG. 1. Absorption of energy by the leaves of a plant averages nearly 90% of solar radiation in the spectral range from ultraviolet through visible frequencies. At near-infrared frequencies, however, absorption falls off sharply and stays at a low level throughout the range at which solar radiation is the most intense. The curve shows absorption by a leaf of the poplar *Populus deltoides*. The divisions of the horizontal scale are proportional to the frequency of the radiation (Gates, 1965).

Gates *et al.* (1964) studied the leaf temperature of several plant species in growth chambers. They found that most plants were cooler than the air between air temperatures of 30 and 40°C. Below about 28°C the leaves were warmer than the air. These tests were conducted under approximately 1700 ft-c from fluorescent and incandescent lamps and at 65% r.h. Unfortunately, air velocity was not stated.

Conduction of heat between the air mass and the plant is probably the major factor in the heat balance. The rate of air movement over the plant surface will control conduction to a large extent by removing heated air molecules from the plant surfaces. Plant temperature can be altered several degrees depending upon the difference between leaf and air temperatures and the velocity of the air (Fig. 2).

Transpiration also removes heat from a plant through the process of evaporation (Gates, 1964, 1968). The heat of vaporization of water, approximately 540 cal g^{-1}, can account for relatively large quantities of heat loss from plants under certain conditions. Rapid transpiration actually may cause the leaves to be cooler than the air in the chamber.

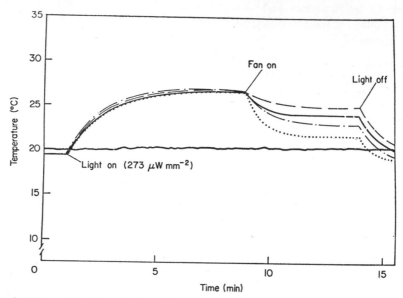

FIG. 2. Effect of wind velocity of 0·6, 1·2, 2·4, and 4·8 m s^{-1} on temperature of illuminated leaf at air temperature of 20 °C and relative humidity of 60%. (VP — 10·5 mmHg).

     – – – – Wind velocity; 0·6 m s^{-1}
     ——— Wind velocity; 1·2 m s^{-1}
     –· – ·· Wind velocity; 2·4 m s^{-1}
     ······ Wind velocity; 4·8 m s^{-1}
     ——— Air temperature
(Matsui and Eguchi, 1972).

The amount of heat lost through evaporation will depend upon the amount of water that is removed from the plant system. This water loss is primarily through leaf stomata and usually to only a small extent through the cuticle of the leaves and the lenticels of the stem. The rate of water loss from a plant is controlled by the rate of diffusion which in turn depends upon several factors. These include the vapor pressure gradient between the internal and external atmosphere and the resistance of the plant surface to the movements of water molecules. The resistance varies from species to species due to the number and position of the stomata and from time to time for individual plants due to the degree of stomatal closure.

Changes in the vapor pressure of the external atmosphere can change the rate of water transfer and thus heat loss. In addition to the vapor pressure of the atmosphere *per se*, air velocity plays an important role.

The reduced temperature (see in Fig. 2) is the result of two factors: the increase in the conduction rate of heat as mentioned previously and the increased loss of water vapor. The marked effect of radiant energy on leaf temperature is shown by the abrupt changes in leaf temperature when the light (273 μW mm^{-2}) was turned on and off (Figs 2 and 3).

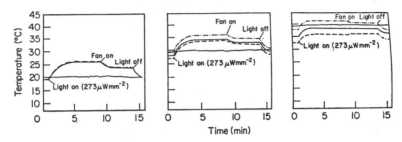

FIG. 3a. Effect of wind velocity of 1·2 m s^{-1} on temperature of illuminated leaf at air temperature of 20 °C under respective conditions of relative humidity of 40, 60, and 80%.

– – – – Leaf temperature at r.h. 40% (VP = 7·0 mmHg)
———— Leaf temperature at r.h. 60% (VP = 10·5 mmHg)
-- - -- Leaf temperature at r.h. 80% (VP = 14·0 mmHg)
———— Air temperature

FIG. 3b. Effect of wind velocity of 1·2 m s^{-1} on temperature of illuminated leaf at air temperature of 30 °C under respective conditions of relative humidity of 40, 60, and 80%.

– – – – Leaf temperature at r.h. 40% (VP = 12·7 mmHg)
———— Leaf temperature at r.h. 60% (VP = 19·9 mmHg)
-- - -- Leaf temperature at r.h. 80% (VP = 25·5 mmHg)
———— Air temperature

FIG. 3c. Effect of wind velocity of 1·2 m s^{-1} on temperature of illuminated leaf at air temperature of 40 °C under respective conditions of relative humidity of 40, 60, and 80%.

– – – – Leaf temperature at r.h. 40% (VP = 22·1 mmHg)
———— Leaf temperature at r.h. 60% (VP = 33·2 mmHg)
-- - -- Leaf temperature at r.h. 80% (VP = 44·3 mmHg)
———— Air temperature
(Matsui and Eguchi, 1972).

The vapor pressure gradient from the stomatal cavity to the outside air decreases rapidly under still air conditions that allow the saturated air to accumulate on the leaf surface. The thickness of this boundary layer determines the rate of water vapor diffusion from the leaf surface. Air movement sweeps the saturated air away, reducing the thickness of the boundary layer and maintaining a steep pressure gradient. In Fig. 2 the drop in temperature when the fan is started is due to an increased transpiration rate. A part of this decrease in temperature is caused by an increase in the rate of removal of heat by conduction, as discussed in a previous paragraph. However, while air movement will remove the water-saturated air from the leaf surface, if the air velocity becomes too great it will cause the closing of the stomata of some plants and thus decrease the loss of water and in turn the loss of heat. Thus, the air flow pattern can have an effect upon the temperature and water use by plants even within a single chamber.

Matsui and Eguchi (1972) also studied leaf temperature at different atmospheric temperatures as affected by the amount of moisture in the air (Fig. 3). The most pronounced atmospheric vapor pressure effect was evident when the atmospheric temperature was high: 40°C. At 20°C there was no readily measurable effect of atmospheric moisture between 40 and 80% r.h. It should be noted that the difference in vapor pressure between 40 and 80% r.h. at 20°C is 7 mmHg compared to a difference of 22·2 mmHg at 40°C.

If radiation, transpiration and conduction–convection are inadequate to balance the heat input to a plant the temperature rises, as shown by the rise in leaf temperature when the light is turned on (Figs 2 and 3). As the temperature rises the amount of radiation from the plant increases rapidly until a new balance is reached at a higher temperature, as shown by the plateau in the graphs. Radiation rises as the fourth power of the absolute temperature and at 30°C or 303 K (absolute) the calories lost through radiation would be almost 114% of those lost at 20°C or 293 K. This rapid change in the rate of heat loss with changes in plant temperature is an important factor in maintaining the leaf temperature within a non-lethal range, both at the upper and lower limits.

A temperature sensor, either shielded or exposed, will seldom be at the same temperature as a plant in a lighted growth cabinet. This is due to the differences in the absorptive and reradiation characteristics of the two bodies in addition to the transpiration effect on the temperature of the plant. However, rapid air movement over the leaves will tend to lower the plant temperature and bring it closer to air temperature as measured by a shielded, aspirated sensor (Matsui and Eguchi, 1972).

The ideal would be to measure plant temperature directly. Unfortunately, a plant does not have a uniform temperature: a shaded portion of a leaf will be at a different temperature from the portion in direct light. Even the angle of the leaf relative to the light source affects the temperature. In addition measuring the temperature of a leaf is difficult. Very small thermocouples or thermistors need to be used and these have small diameter leads that are easily broken. The mass of the unit and the leads can absorb or conduct an appreciable amount of the heat away from or into a thin leaf and thus actually cool or heat the leaf at the point of measure-ment. Radiometers that determine the temperature by measuring the amount of heat being radiated from a surface are unsatisfactory for indi-vidual plants or small groups of plants in a chamber. This is due to the irregularity of the plant surface and interference from the chamber, the instrument itself and the presence of the user.

Temperature Control Over Plant Growth

Temperature is the most frequently measured of all the environmental factors when studying plants in the field or in growth chambers. Tempera-ture control over plant growth is manifest in many ways. These include total plant growth, a degree of deviation from "normal" growth, breaking of bud dormancy and flowering. However, it is important to remember that temperature operates through changes induced in the internal processes of the plant.

Plants respond to temperature in a complex buffered manner. This is due to several factors. First, the thin, flat shape of a leaf makes it an efficient conductor of heat. Nevertheless the specific heat is high due to the water content and the leaf changes temperature much more slowly than the air around it. Second, temperature controls plant growth and development by modulating the rate of the numerous physical and biochemical processes that are a part of the physiology of a plant.

PHYSICAL PROCESSES

The physical processes of diffusion and transpiration are directly affected by temperature. An increase in temperature increases the kinetic energy of the molecules and therefore increases their partial pressures. The rate of diffusion of both organic and inorganic molecules within the aqueous systems of a plant, both internal and external to the cell membranes, is an important factor in the rate of growth. Likewise, a change in the vapor pressure of water in the leaf is a major factor in controlling the rate of water loss of a plant and in turn the rate of the mineral movement via the transpiration steam from the roots to the aerial portions.

METABOLIC PROCESSES

Photosynthesis involves light reactions which are of course not tempera-ture dependent, but the products of the light reactions are utilized in dark reactions which are temperature sensitive. Thus, the temperature-dependent phases may limit the rate of the overall photosynthetic process.

Respiration and other metabolic processes consist of a series of linked reactions, each of which is affected by temperature. The different biochemical reactions that occur in a plant have different maximum, minimum and optimum temperatures. The effects on development are most pronounced when temperature is near the minimum or maximum for plant growth. As the lower limit of temperature is approached there is usually a gradual decrease in the growth rate which often results in stunted plants with thick leaves. There are exceptions, such as redwood seedlings where a decrease from 15 to 11°C night temperature caused an abrupt decrease in height growth and dry matter production of 70–80% (Hellmers, 1966). For most plants growth ceases between 0 and 10°C but some processes such as photosynthesis and respiration can continue very slowly at much lower temperatures (Kozlowski and Keller, 1966; Kramer and Kozlowski, 1960; Pharis et al., 1970). As the maximum cardinal temperature is approached there is often a rapid increase in stem elongation with the production of thin leaves and a decrease in total dry matter. There is usually an abrupt decrease in growth between the temperature that produces extreme elongation and that which is too high for plant growth and survival.

Shirley (1936) studied high temperature effects on seedlings of several coniferous species. Resistance to being killed by excessive heat for short periods was increased with increased plant age owing to the greater mass of tissue. Tops, normally being repeatedly subjected to rapid temperature changes, developed a greater resistance to excessive heat than did the roots. For instance, a temperature of 44·3°C for 5 h did not cause severe damage to the tops but was lethal to roots.

There is some indication that there are physiological reaction discontinuities at or near temperatures of 15, 30, and 45°C. The evidence is reviewed in detail by Nishiyama (1972). The effect appears to be related to changes in the physical properties of water at these temperatures, which in turn affect various physiological processes (Drost-Hansen, 1966, 1969) and the bioelectric potential of cells (Thorhaug, 1971). Two examples cited by Nishiyama are shown in Figs 4 and 5.

Enzymatic Reactions. The temperature effect on chemical reactions is usually expressed in terms of Q_{10} which is the ratio of the rate of the reaction between two temperatures that differ by 10°C. Inorganic chemical reactions tend to have a Q_{10} of about 2 which means they double each time the temperature is raised 10°C. For enzymatically controlled reactions the Q_{10} frequently approaches 2 between 10 and 30°C.

Most enzymatic reactions slow down markedly below 10°C and for all

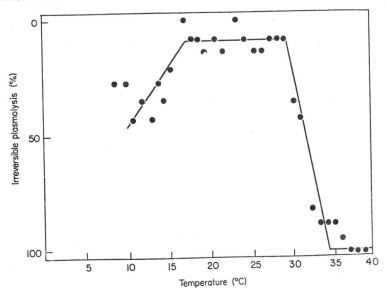

Fig. 4. Percentages of *Nitella* cells surviving 3 days' exposure to temperatures indicated. Data by Thorhaug from Drost-Hansen (1968).

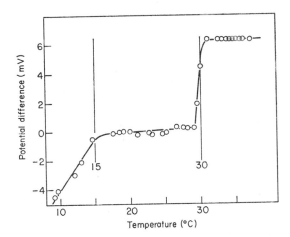

Fig. 5. Changes of mean potential (closed dots) from resting potential at 25 °C and standard error of mean potential (vertical lines) across the protoplasm of thirty *V. ventricosa* cells *vs* temperature ($E = \psi i - \psi o$). (Thorhaug, 1971).

practical purposes cease a few degrees below 0°C. These effects coincide
with changes in the physical properties of water, the media in which
biochemical reactions occur in the living cell. The viscosity of water
increases rapidly as the temperature drops below 10°C. Maximum density
occurs at about 4°C and below this temperature the water tends toward
crystalization. Cell sap will differ in its physical properties from those of
pure water due to dissolved organic and inorganic substances. As the
temperature rises above 30°C most enzymatically-controlled reactions also
tend to slow and stop. Excessive heat finally denatures the protein of the
enzyme, a condition that results in a change in the configuration of the
protein molecule. This changes the spatial relation or causes a destruc-
tion of the reaction sites that are required to bind the substrates to the
enzyme for the reactions. The denaturing of proteins by heat involves a
time constant: a time–temperature interaction. Consequently, heat
damage may be caused either by a high temperature for a short period or
a lower temperature over a longer time.

The growth process consists of a series of reactions, most of which are
enzymatically controlled. Rapid changes in rates of reaction for different
enzymatic systems range from freezing point to temperatures above 60°C
(Massey et al., 1966). Therefore, increasing the temperature may increase
the rate of one enzymatic reaction in the plant system while the same
change in temperature may increase at a different rate or may even slow
down another reaction. In addition, the metabolic pathways in a plant often
include more than one route from photosynthate to storage product or
structural molecule. Thus, the change in temperature may cause only a
shift from, or to, a more or less efficient pathway. Nevertheless, many
plants grow most rapidly at temperatures between 10 and 30°C. Beyond
this range of temperature, in either direction, the reaction rates decrease
rapidly. Langridge (1963), in reviewing biochemical activities in response
to temperature, points out that at temperature extremes the rate of growth
may be limited by the velocity of a single reaction.

STRATIFICATION

Seeds of certain plants are dormant under normally favorable growing
conditions and require a cold (0–6°C) moist treatment before they will
germinate. This cold, moist treatment is termed *stratification*. Dry, cold
storage will not satisfy the cold requirement. The length of the treatment
required varies with species. The pines and birches may require from 15 to
90 days of stratification before planting. In general, but not always, the
more southern the species the less stratification time is needed. The length
of time seeds are kept in stratification and the temperature of the treatment

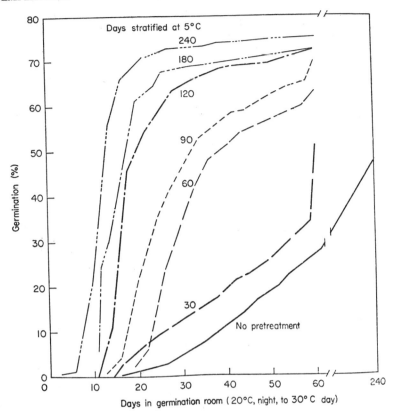

FIG. 6. Effect of stratification period on germination of balsam fir (*Abies balsamea*) seeds. (U.S. Forest Service, 1948)

can affect not only their total germination percentage but also the time required to obtain germination (Fig. 6). Recommended duration of stratification at 5°C for seeds of some tree species are given in Table I.

TEMPERATURE *per se*

The growth rate, in common with the enzymatic reactions that control it, exhibits an increase that is almost linear for most plants as the temperature is increased from 10 to 30°C. The response varies with the species on a time–temperature basis. Some plants respond to total heat units over a 24 h period (Fig. 7), others respond primarily to day temperature or to night temperature (Fig. 8) and a third group require a day–night temperature differential called "thermoperiod" (Kramer, 1957; Went, 1957).

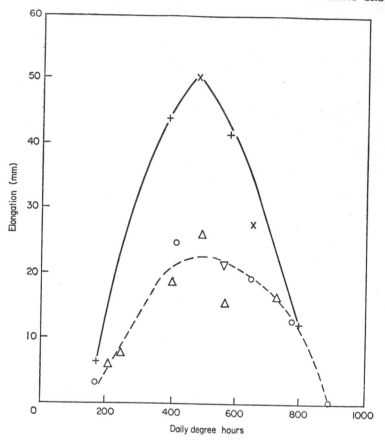

FIG. 7. Height growth response of Eastern Hemlock seedlings and 1 year-old plants to total heat units given over 24 h. The distribution of the heat units between day and night had little effect. (Drawn from data presented by Olson *et al.*, 1959.)

The physiological processes that come into effect and that cause these differences have not been determined.

The effect that temperature can have on some of the different parameters of growth is evident from many past studies. For instance, high night temperature has been observed to stimulate height growth at the expense of dry matter production (Fig. 8). Loblolly pine which produces from one to seven flushes of growth during the year, depending upon its age and location, exhibits an inverse relationship with temperature: flush number is increased and flush length shortened with increased temperature (Fig. 9). Thus the trees grow to approximately the same height under a wide

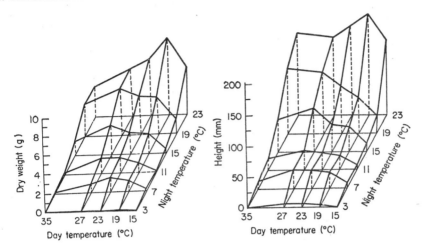

FIG. 8. Average dry weight and height of thirty-six Engelmann spruce seedlings from each of thirty combinations of day and night temperature where the plants were grown for 24 weeks. (Hellmers *et al.*, 1970).

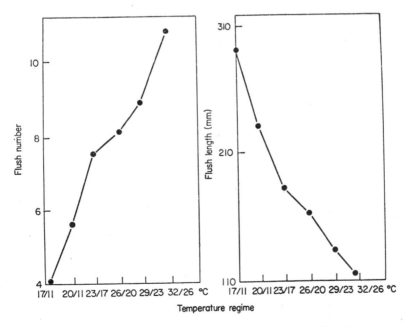

FIG. 9. Mean node number and mean internode length of loblolly pine trees grown under various day/night temperatures for 52 weeks (Mulroy, 1972).

range of temperatures. In a study using wheat Friend and Pomeroy (1970) found a response in leaf length and cell number that had different temperature optima (Fig. 10).

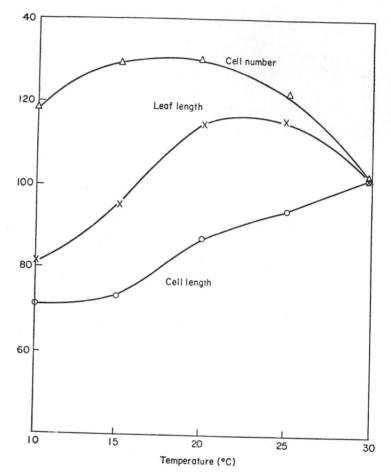

Fig. 10. Relative changes in lamina length and epidermal cell length and number in Marquis spring wheat grown at different temperatures. Points are means from twelve plants (Friend and Pomeroy, 1970).

Increased or decreased production of certain products in plants is often evident under extremes within the non-lethal range of temperature. For example the production of the essential oil that gives the strawberry its taste is controlled by temperature (Went, 1957). Also, anthocyanins which cause the color in many plant flowers and leaves are temperature responsive; usually, the cooler the temperature the higher the concentration of the pigment, as is shown in *Begonia* leaves and flowers (Fig. 11) (Martin *et al.*, 1972).

A temperature of 10°C as compared with 25°C was found to increase the level of almost all free amino acids in a variety of plants (Taylor *et al.*, 1971). It was also found that the lower temperature delayed the loss of starch for 24 h.

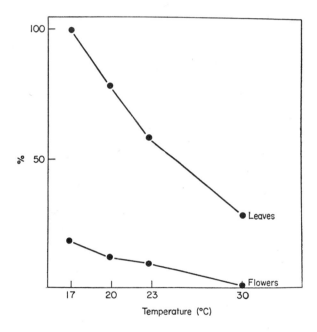

FIG. 11. Percent of anthocyanins of leaves and flowers of *Begonia gracilis* var. Carmen. The leaf content at 17 °C is equated to 100% anthocyanin content (Martin *et al.*, 1972).

Many types of processes are involved in the growth of a plant. Therefore, it is to be expected that temperature requirements differ not only between species but during different stages of growth and development in the life cycle of a plant. For example, cotton has a lower optimum temperature for flower production and flower set than for vegetation growth (Hesketh and Low, 1968; Mauney 1966).

TABLE I

Stratification time at 5 °C required to obtain rapid seed germination for representative study species (data from U.S. Forest Service 1948)

Stratification at 5 °C	
60–90 days	*Pinus echinata*
Abies amabilis	*P. elliottii*
A. concolor	*P. strobus*
A. magnifica	*Quercus rubra*
A. procera	*Q. macrocarpa*
Chamaecyparis nootkatensis	*Q. falcata*
Cupressus arizonica	
Libocedrus decurrens	30–60 days
Picea glauca	*Larix laricina*
P. sitchensis	*Picea mariana*
Pinus lambertiana	*Pseudotsuga menziesii*
P. monticola	*Taxodium distichum*
Tsuga heterophylla	*Thuja occidentalis*
Acer negundro	*T. plicata*
A. rubrum	*Alnus rubra*
A. saccharum	*Betula alleghaniensis*
Betula populifolia	*Quercus velutina*
Carya illinoensis	*Q. coccinea*
Celtis occidentalis	
Fagus grandifolia	90–150 days
Fraxinus americana	*Carya ovata*
F. pennsylvanica	*C. laciniosa*
Nyssa sylvatica	*C. tomentosa*
Ulmus americana	*C. glabra*
U. rubra	*Gymnocladus dioicus*
	Juglans cinerea
30–45 days	*Juglans nigra*
Abies grandis	*Prunus serotina*
Chamaecyparis lawsoniana	*Tilia americana*
Larix occidentalis	*Tsuga canadensis*
Picea rubens	

THERMOPERIODISM

Thermoperiodism is defined as the response of plants to a periodically fluctuating temperature such as the differences between day and night temperatures. The phenomenon of thermoperiodism as a controlling factor in plant growth was first described by F. W. Went (1944). A periodic fluctuation in temperature is required for good plant growth by some but not all plant species. Many plants, such as cocklebur, will grow well when the day and night temperatures are identical.

Thermoperiodism may control flowering as well as vegetative growth. For example, optimum flowering of tomato plants occurs under the conditions of a 25°C day temperature with a 15°C night temperature. No specific region of perception or hormone is known to control the plant's response to thermoperiod. It is thought that the thermoperiod effect is related to the accumulation of photosynthate and storage during the day and respiratory utilization of these products by night. Night temperatures as high as day temperatures (26°C) are lethal for some desert plants such as *Baeria chrysostoma* (Went, 1957). They can only survive if they have a thermoperiod with a low night temperature. Studies have shown changes in top-to-root ratios of plants with differences in thermoperiod. Thermo-periods need not always give a warmer day than night condition to obtain optimum growth. For example some coniferous seedlings seem to require a thermoperiod but the day and night temperatures can be reversed to produce equal growth (Hellmers, 1963).

Temperature and thermoperiod during the time of seed maturation and germination can effect the light requirement some seeds have for germination (Table II).

TABLE II

Effect of seed maturation temperature on the by-passing of Coker 319 tobacco seed germination light requirement by an alternating germination temperature. Germination (expressed as a percentage) after 0–4 cycles of thermoperiod in the dark or light treatment at (25) °C

Daily cycles at 25/15 °C (dark)	Maturation temperature % germination		
	30/26	26/22	22/18
0	0	2	12
1	0	14	63
2	20	58	81
3	77	93	96
4	91	100	97
Light control	98	99	99

VERNALIZATION

Biennial plants require a cold treatment to change from the vegetative to the reproductive phase. Some perennial plants require a cold treatment to break the dormancy of vegetative buds as well as to form flower buds and to induce their growth. The cold treatment, termed vernalization, involves temperatures between 0 and 6°C and a time factor. Even though some plants do not have vernalization requirements for flowering, the process can be hastened by a vernalization treatment. Thus winter rye through vernalization can be induced to flower as rapidly as spring varieties. Carrots and beets are biennial and require a cold treatment for flowering.

Similar to other temperature requirements, the amount of chilling needed for vernalization varies with species and even varieties within species. For example, Concord grapes require 3500 h below 7°C to grow and bear fruit, and Mayflower peach trees require 1150 h, while Red Ceylon peaches require only 50 h at 7°C. An extensive list giving the vernalization requirement for flowering of a large number of angiosperms is published in Table 16 of the *Handbook of Biological Data* (Spector, 1956).

Woody perennial plants of the temperature zone such as oak often require a vernalization treatment once buds are set if growth is to resume. Other plants including sweetgum and loblolly pine can repeatedly break bud and grow under a long photoperiod but not under a short photoperiod of 10 h or less (Hellmers and Hesketh, 1973).

Bud set itself may be temperature dependent as well as photoperiodic dependent.

Root Temperature

Temperature affects the growth, anatomy and metabolism of roots (Nightingale, 1935). Root temperature exerts control over plant growth by affecting the uptake of water and minerals and by affecting initiation and root growth. As with temperature effects on top growth, a temperature–time relation exists for root growth (Carpenter *et al.*, 1973). In greenhouses and chambers care should be taken to use water or nutrient solution that is at a temperature that does not adversely affect root development. In each of the temperature-controlled greenhouse plenums at the Duke University phytotron we have 60 gal stainless steel holding tanks included in the nutrient solution and demineralized water systems. This allows the contents to approach the temperature of the room between periods of use and prevents a temperature shock, especially in the 32 and 17°C houses. As yet we have not installed these in the chamber plenums.

What Temperature to Use in a Study

Temperature or temperatures to be selected for use in a study will depend on whether the objective is to study temperature effects *per se* or to study some other factor while maintaining a favorable temperature regime.

Where the objective is to study temperature effects, then the limits of temperature to be used will be governed by the anticipated use of the results. Whether it is the growth potential of the plant, or the suitability of the plant for use in an area that is to be determined, will make a difference in the range of temperatures needed. In studying the potential of a plant in relation to temperature a range of conditions that extend from almost freezing to upper lethal temperatures should be used. Plants can have a greater potential for growth than is shown by specimens growing on their native sites. Two examples are Engelmann spruce and radiata pine. The former is capable of growing under night conditions at a much faster rate than occurs within its natural range (Hellmers *et al.*, 1970). The latter species grows better under the cooler temperature conditions found in New Zealand than it does within its natural range on the coast of California (Hellmers and Rook, 1973).

Screening plants to introduce to a new climatic region can be a more complicated process. Not only is growth of the plant of interest but ability to survive, flower and set seeds can be of major importance. This then involves not only a range of temperatures based on the average temperatures for an area but also extremes of temperatures and their duration in relation to the physiological conditions of the plant (Shirley, 1936). A critical temperature during the seedling stage can be quite different from a critical temperature during vegetative growth of a more mature plant. Also during the time of flowering many plants are extremely sensitive to temperature and an unfavorable temperature for a very short time can drastically reduce seed set.

Many plants will grow reasonably well over a wide range of temperatures. Therefore, if the objective is to maintain a favorable temperature while studying the effects of another environmental factor the investigator has some latitude within which to work. Nevertheless, because temperature interacts with so many processes it is advisable to check the response obtained by using several temperature conditions, including alternating day and night temperatures as well as different constant ones.

Root temperatures can be very important, as described in the previous section. Incorporation of them into a study causes many additional complications but can yield much valuable information in matching plants to soil types. Root temperature effects on plant growth processes are in need of much more investigation.

Each researcher will have to decide for himself for each particular study the extent to which temperature effects should be investigated or controlled to provide the necessary refinement of results. Temperature controls every biochemical process and interacts with all the other environmental factors in controlling the various plant growth processes. Thus, temperature selection must be based on all available information and be given extensive thought when planning any meaningful plant growth experiment.

References

Carpenter, W. J., Hansen, E. N. and Carlson, W. H. (1973). Medium temperatures effect on geranium and poinsettia root initiation and elongation. *Journal of the American Society of Horticultural Science* **98**, 64–66.

Drost-Hansen, W. (1966). The effects on biological systems of high-order phase transitions in water. *Annals of the New York Academy of Science* **125**, 471–501.

Drost-Hansen, W. (1968). Role of water structure in various membrane systems. Paper for Coral Gables Conference on Physical Principles of Biological Membranes.

Drost-Hansen, W. (1969). Structure of water near solid interfaces. *Industrial and Engineering Chemistry* **61**, 10–47.

Drost-Hansen, W. and Thorhaug, A. (1967). Temperature effects in membrane phenomena. *Nature* **215**, 506–508.

Friend, D. J. C. and Pomeroy, M. E. (1970). Changes in cell size and number associated with the effects of light intensity and temperature on the leaf morphology of Wheat. *Canadian Journal of Botany* **48**, 85–90.

Funk, J. P. (1963). Improvements in polythene-shielded net radiometers. Engineering Aspects of Environment Control for Plant Growth, C.S.I.R.O., pp. 248–256, Australia.

Gates, D. M. (1964). Leaf temperature and transpiration. *Agronomy Journal* **56**, 273–277.

Gates, D. M. (1965). Heat transfer in plants. *Scientific American* **213**, 76–84.

Gates, D. M. (1968). Transpiration and leaf temperature. *Annual Review of Plant Physiology* **19**, 211–238.

Gates, D. M., Heisey, W. M., Milner, H. W., and Nobs M. A. (1964). Temperatures of *Mimulus* leaves in natural environments and in a controlled chamber. *Annual Report Carnegie Institute of Washington Yearbook* **63**, 418–430.

Hellmers, H. (1962). Temperature effect on optimum tree growth. *In* "Tree Growth." (T. T. Kozlowski, ed.), pp. 275–287. Ronald Press, New York.

Hellmers, H. (1963). Some temperature and light effects in the growth of Jeffrey pine seedlings. *Forest Science* **9**, 189–201.

Hellmers, H. (1966). Growth response of redwood seedlings to thermoperiodism. *Forest Science* **12**, 276–283.

Hellmers, H., Genthe, M. K. and Ronco, F. (1970). Temperature affects growth and development of Engelmann Spruce. *Forest Science* **16**, 447–452.

Hellmers, H. and Hesketh, J. D. (1974). Phytotronics and modeling plant growth. *In* "Mechanisms of Regulation of Plant Growth", pp. 637–644. Royal Society of New Zealand, Wellington, New Zealand.

Hellmers, H. and Rook, D. A. (1973). Air temperature and growth of radiata pine seedlings. *New Zealand Journal of Forestry Science* **3**, 271–285.

Hesketh, J. D. and Low, A. (1968). Effect of temperature on components of yield, and fiber quality of cotton varieties of diverse origin. *Cotton Growers Review* **45**, 243–257.

Kozlowski, T. T. and Keller, T. (1966). Food relations of woody plants. *Botanical Review* **32**, 293–382.

Kramer, P. J. (1957). Thermoperiodism in trees. *In* "The Physiology of Forest Trees". (K. V. Thimann, ed.), pp. 573–580. Ronald Press, New York.

Kramer, P. J. and Kozlowski, T. T. (1960). "Pyhsiology of Forest Trees." McGraw-Hill, New York.

Langridge, J. (1963). Biochemical aspects of temperature response. *Annual Review of Plant Physiology* **14**, 441–462.

Martin, C., Paynot, M. and Vallee, J. C. (1972). Quelques effets de la temperature sur la floraison le metabolisme amine et l'anthocyanogenese. *In* "Phytotronique et Prospective horticule" (P. Chouard and N. de Bilderling, eds). Gauthier-Villars, Paris.

Massey, V., Curti, and Ganther, H. (1966) A temperature dependent contermational change in D-amino acid oxidase and its effect on catalysis. *Journal of Biological Chemistry* **241**, 2347–2357.

Matsui, T. and Eguchi, H. (1972). Effects of environmental factors on leaf temperature in a temperature controlled room. *Environment Control in Biology* **10**, 15–18.

Mauney, J. R. (1966). Floral initiation of upland cotton, *Gossypium hirsutum* L. in response to temperature. *Journal of Experimental Botany* **17**, 452–459.

Mulroy, J. (1972). Some effects of temperature on growth and photosynthesis in Loblolly pine (*Pinus taeda* L.) seedlings. MA. Thesis, Duke University, Durham, North Carolina.

Nightingale, G. T. (1935). Effects of temperature on growth, anatomy and metabolism of apple and peach roots. *Botanical Gazette* **96**, 581–639.

Nishiyama, I. (1972). Physiological kinks around 15, 30 and 45°C in plants. Hokkaido National Agricultural Experiment Station, Bulletin 102, pp. 125–167.

Olson, J. S., Stearns, F. W. and Nienstaedt, H. (1959). Eastern hemlock seeds and seedling response to photoperiod and temperature. *Connecticut Agricultural Experimental Station Bulletin*, 620.

Pharis, R. P., Hellmers, H. and Schuurmans, E. (1970). Effects of subfreezing temperature on photosynthesis of evergreen conifers under controlled environment conditions. *Photosynthetica* **4**, 273–279.

Read, W. R., Cunliffe, D. W., Chapman, H. L. and Kowalczewski, J. J. (1963). Naturally lit plant growth cabinets, pp. 102–122. *In* "Engineering Aspects of Environment Control for Plant Growth". C.S.I.R.O., Australia.

Shirley, H. L. (1936). Lethal high temperatures for conifers and the cooling effect of transpiration. *Journal of Agricultural Research* **53**, 239–258.

Spector, W. S. (ed.). (1956). Handbook of Biological Data. W. B. Saunders, Philadelphia.

Sutton, D. J. and McNall, P. E. Jr. (1964). Two-sphere radiometer. *Heating, Piping, and Air-Conditioning* **26,** 157–162.

Taylor, A. O., Jepson, N. M. and Christeller, J. T. (1971). Plants under climate stress. III. Low temperature high light effects on photosynthetic products. *Plant Physiology* **49,** 798–802.

Thorhaug, A. (1971). Temperature effects on *Valonia* bioelectric potential. *Biochemica et Biophysica Acta* **255,** 151–158.

U.S. Forest Service. (1948). Woody-Plant Seed Manual. U.S.D.A. Miscellaneous Publications 654.

Went, F. W. (1944). Plant growth under controlled conditions, II. Thermoperiodicity in growth and fruiting of the tomato. *American Journal of Botany* **31,** 135–140.

Went, W. (1957). "The Experimental Control of Plant Growth." Chronica Botanica Co., Waltham, Mass.

CHAPTER III

Light

There is little merit in entering into a detailed discussion of the physical characteristics of light since the subject has been thoroughly discussed many times (Allphin, 1959; Elenbaas, 1959; Seliger and McElroy, 1965; IES, 1968; Clayton, 1970; Bickford and Dunn, 1972), The main point to reiterate is that light is energy. When a plant part absorbs that energy it may be (1) converted into heat; (2) reradiated as fluorescence; (3) utilized to sensitize a reaction which then becomes susceptible to other kinds of light or to accelerate chemical reactions that would otherwise take place very slowly; (4) used as the primary energy source to run an initial reaction upon which many subsequent reactions depend.

Perhaps the two most important rules of photoreactions are the Grotthus-Draper law which states that only light that is absorbed can be used in producing a chemical change, and the rule that states that every molecule taking part in a photochemical reaction absorbs one quantum of the radiation: the Stark-Einstein rule of photochemical equivalence. Since the main interest in light as an environmental factor is in the way it operates the photobiological systems, we are principally interested in the number of quanta absorbed per unit area per unit time within the absorption bands of the various biological pigment systems.

Light by definition is that portion of the electromagnetic spectrum visible to the human eye. Current definitions suggest a wavelength range of from 380 to 760 nm. The total electromagnetic spectrum exceeds the light portion in both directions, extending from cosmic rays with a wavelength of about 1×10^{-5} nm to the 1×10^8 m electric waves of power transmission. Although a gradual transition takes place from one portion of the spectrum to another, the different regions are usually referred to as if they were sharply defined categories. Each of the various categories may have a direct or indirect effect on biological organisms. The germicidal effects at 250 nm and the erythemal reaction to 290–300 nm radiation are examples. However, these effects are in the ultraviolet and not pertinent to this discussion which is concerned with visible light.

In order to discuss light as an environmental factor we must first determine how to measure it. Since we use artificial light sources in controlled-environment chambers it might prove quite useful to know how artificial sources generate light, which in turn would illustrate how the spectral distribution of the emitted energy depends on the method of generation.

Measurement

Light is most often measured in terms of illuminance, defined as the luminous flux per unit area. The unit of illuminance is the lux ($= \text{lm m}^{-2}$) or at high light levels the hectolux (hlx). In the USA, light meters are

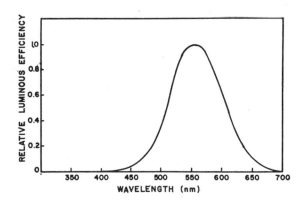

FIG. 1. Relative spectral luminous efficiency for photopic vision.

calibrated in terms of lm ft^{-2} or foot candles (ft-c), but conversion to lux is relatively easy (1 ft-c = 10·764 lx). Luminous flux (lumen) is therefore a method of *visually* evaluating radiant energy and is accomplished by multiplying the energy at each wavelength by the efficiency of photopic vision for that wavelength; with the energy at maximum sensitivity, 555 nm, equal to one. The result is a spectral luminous efficiency curve based on values standardized by the Commission Internationale de l'Eclairage (C.I.E.) (Fig. 1). Detectors of the type used in light meters are usually equipped with filters to provide a spectral response as nearly like the standard curve as possible.

Illumination meters have a reputation of inaccuracy especially at high light levels. Where a barrier layer cell and microammeter are used, the meter alone can have an error of $\pm 8\%$ at any point above $\frac{1}{4}$ full scale. Experience has shown that it is not unusual for two apparently identical meters to disagree by as much as 30% at illuminance levels of several hundred hectolux. However, a recent comparison of four light measuring systems used in the NCSU phytotron shows that with reasonable care the variation between instruments can be held to less than 10% (Table I).

TABLE I

Comparison of various illumination meters for monitoring con- trolled-environment room light.

	Illuminance (hlx)		Fluorescent + Incandescent
Meter	Fluorescent	Incandescent	
Weston	419	43	430
G. E.	425	43	441
Gamma Sci.	436	42	468
Lambda Inst.	425	45	460

Illuminance measurements for any biological response other than photo- pic vision are obviously not correct, because other responses do not have the same spectral sensitivity. However, when used as a relative indication of energy at various distances, at different times or between the same kinds of light systems the measurement is useful. Needless to say, the data would lose validity if different light sources or combinations of light sources were to be measured. In other words, 300 lx from incandescent lamps would not induce the same plant response as 300 lx of fluorescent light although the same visual stimulation would be obtained.

Many investigators have criticized illuminance as meaningless for plant studies and recommended total energy measurements as the only valid system. Unfortunately, absolute energy measurements as W m^{-2}, langleys min^{-1}, etc. are usually made with instruments that are equally or at least partially responsive to all wavelengths; including the infrared region out as far as 66 μm. Absolute energy measurements therefore suffer from the same shortcomings as illumination data; they are not relevant to plant response except in a relative way. On an absolute basis energy measurements of some light sources such as the sun or incandescent lamps will seem unusually high because of the large emission of long wavelength radiation. Correspondingly, total energy from fluorescent lamps may seem low because they do not emit much infrared. Silicon cell pyranometers are often calibrated for sunlight against an Eppley thermopile type of instrument. While these instruments read out in absolute energy units they do not have a flat response. Since the response is spectrally sensitive, errors will result when measurements are made under light sources for which the instrument has not been calibrated. As a result, the irradiance data given in Table II for controlled-environment rooms is not as exact as that for sunshine in Table III.

TABLE II

Energy relationships 1·5 m from the lamps in a fluorescent–incandescent lighted 1·22 × 2·44 m controlled-environment room

	Fluorescent	Incandescent	Fluorescent + Incandescent
Installed watts	6020	2400	8420
Hectolux	425	45	470
Microeinsteins m^{-2} sec^{-1}	670	97	770
Nanoeinsteins cm^{-2} sec^{-1}	67·0	9·7	77·0
Watts m^{-2}	91·5	162	250
Milliwatts cm^{-2}	9·2	16·2	25·0
Langleys min^{-1}	0·14	0·20	0·34

TABLE III

Energy relationships inside an unshaded glasshouse at 2 p.m. during October at Raleigh, N. C.

	Heavy overcast	Full Sun Horizontal to surface	Full Sun Perpendicular to sun
Hectolux	17	800	1000
Microeinsteins m^{-2} sec^{-1}	40	1625	2000
Nanoeinsteins cm^{-2} sec^{-1}	4·0	162·5	200·0
Watts m^{-2}	12·5	5900	9600
Milliwatts cm^{-2}	1·25	590	960
Langleys min^{-1}	0·025	0·825	—

W. W. Biggs, (Lambda Instrument Corporation, personal communications) tells us that a silicon cell sensor will deviate from an ideal response in the following way when measurements are made under sources with different spectral distributions. The ratio of sensor output in light being measured, Q, to the output in the calibration source Q_c is $Q/Q_c =$ $\dfrac{E_m \int r(\lambda)\, e(\lambda)\, d\lambda}{E_{mc} \int r(\lambda)\, e_c(\lambda)\, d\lambda}$ where E_m is the maximum amplitude of $E(\lambda)$, $e(\lambda)$ the normalized spectral distribution and $r(\lambda)$ the normalized response.

The same relationship holds for a sensor with an ideal response (i) calibrated under the same standard source

$$\frac{Q_i}{Q_{ic}} = \frac{E_m \int r_i(\lambda)\, e(\lambda)\, d\lambda}{E_{cm} \int r_i(\lambda)\, e_c(\lambda)\, d\lambda}$$

Combining the two equations

$$\frac{Q/Q_c}{Q_i/Q_{ic}} = \frac{\int r(\lambda)\, e(\lambda)\, d\lambda}{\int r(\lambda)\, e_c(\lambda)\, d\lambda} \cdot \frac{\int r_i(\lambda)\, e_c(\lambda)\, d\lambda}{\int r_i(\lambda)\, e(\lambda)\, d\lambda}$$

which is the relative sensor error.

Estimated errors would amount to only 3–4% high with incandescent light but with fluorescent ones readings would be 36–40% low.

Another type of measurement receiving some support is to determine the energy between 400 and 700 nm with an instrument having a flat response (no spectral sensitivity) over that wavelength range. This energy is referred to as photosynthetically active radiation (PAR), and the units used are microeinsteins m^{-2} sec^{-1}. The major objection to PAR measurements is the failure to include the photomorphogenic radiation between 700 and 800 nm which has a major effect on plant growth and development.

Probably the most descriptive measurement, and the one with fewest faults, would be one that includes the spectral energy distribution (SED) over a wide range of wavelengths, at least 400–800 nm. Spectral irradiance describes the light source adequately and the energy in the various effective regions can be quickly determined. Those interested in PAR can easily calculate it from SED curves and the red–far-red ratio that controls the phytochrome system also can be quickly obtained. A number of spectroradiometers are available commercially and while some are clumsy to use and some have poor reproducibility, the main disadvantage is cost.

Tables II and III illustrate the peculiarities of radiant energy measurement. In a controlled environment room (CER) where the fluorescent–incandescent installed watts are 2·5 to 1, the illuminance from the two sources is 9·4 to 1, the PAR radiation 7 to 1 but the total energy is 0·56 to 1. The data therefore suggest that 2400 W of incandescent lamps produce nearly twice as much total energy as 2·5 times as many watts of fluorescent lamps. Nevertheless, few biologists would suggest that the incandescent lamps would produce the best plant growth. In terms of light output the fluorescent system is nearly 4 times more efficient and nearly 3 times more productive of PAR radiation than the incandescent lamps and this relationship is verified by the plant response. Assuming PAR is the best indication of the light for plant growth then total energy measurements of incandescent light (Table II) or from the sun (Table III) result in values 3–3·6 times too high.

Ail light-measuring instruments should give special attention to entrance optics. Measurement of light from broad sources commonly used in plant growth chambers or measuring sun and skylight requires that the receiver follow the cosine law. The cosine law simply states $E_2 = E \cos \theta$; that the illumination varies as the cosine of the angle of incidence. The error due to high angles of incidence, which can amount to 25% or more, is compounded by the fact that the sensor may be set below the rim of the cell mount and partially obscured. Diffusing discs are commonly used to provide a cosine response. However, in many cases the result has been a poor match, and in others the fit to the cosine response has not actually been tested. The cosine correction of the Lambda instrument shown in Fig. 2 is the type of data that should be furnished with each system.

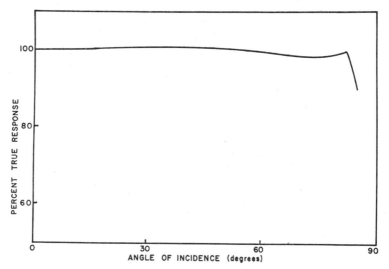

FIG. 2. Cosine correction typical of Lambda quantum and illumination sensors (Lambda Instrument Corporation).

Illumination standards set by the C.I.E. are used by the majority of biological investigators. Irradiance standards unfortunately have not been agreed upon by biologists and every conceivable unit can be found in the biological literature. Solar radiation is most frequently measured as cal cm^{-2} min^{-1} although this is sometimes called a langley min^{-1}. The W m^{-2} is receiving considerable use in plant growth chamber measurements. PAR will receive increasing attention as a valid measurement for plant growth and these units will probably be microeinstein m^{-2} s^{-1}. Thus, the SEPEL growth chambers with a 2·5:1 installed wattage ratio of fluorescent and incandescent lamps would provide an illuminance of 470 hlx. The measured total energy would be about 250 W m^{-2}. PAR radiation would be 152–167 W m^{-2} (calculated) or 770 microeinstein m^{-2} s^{-1} and there would be about 14 W m^{-2} of far-red between 700 and 800 nm.

Whether illuminance, irradiance, spectral irradiance or incident quanta are selected for measuring the radiant flux density, a full description of the light sources should always be included. In this way, other investigators with less elaborate instrumentation can duplicate conditions. It should also be noted that irrespective of the very valid arguments against illuminance data, a large segment of the scientific community can relate light to lux better than to any other value. Lux can be mentally compared to an amount of light recognized by the investigator. This, plus the relatively low cost of the instrumentation, makes it unlikely that illuminance measurements will be replaced in the near future. However, devices like the Lambda instrument are very likely to facilitate a changeover to PAR measurements —especially since the system can also measure illuminance and total energy and the cost is relatively low.

Light Sources

The most common light source used for growing plants is the fluorescent lamp, generally the cool white type. Cool white or warm white lamps have proven to be the best of the many fluorescent lamps for plant growth purposes. Daylight, white and the newer deluxe types are less satisfactory. Special-phosphor lamps like Plant Gro, GroLux and Wide Spectrum GroLux are still controversial in that plant growth is not consistently better than that obtained with cool white. Moreover, the increased cost of the special phosphor lamps usually outweighs whatever improvement is obtained. Plant-growing lamps are still being developed however, and recent improvements are very encouraging.

FLUORESCENT–INCANDESCENT

Irrespective of the type of fluorescent lamp used, the addition of some light from incandescent sources usually improves plant growth. The improvement is smallest with fluorescent lamps like cool white which produce an appreciable amount of red, and greatest in the bluer "daylight" lamps. The amount of incandescent usually added is about 10% of the illuminance or as a rule of thumb about 30% of the installed lamp watts. The value for the added incandescent light was obtained by Parker and Borthwick (1949) in rooms lighted by carbon arc lamps (Table IV) and by Dunn and Went (1959) in fluorescent lamp rooms (Fig. 3). There is

TABLE IV

Yield of Biloxi soybeans after 4 weeks' growth on 16 h days under carbon arc and carbon arc plus incandescent radiation (Parker and Borthwick, 1949)

Per plant		Arc	Arc + Incandescent
Dry weight	g	1·80	2·45
Starch	mg	37	94
Sugars	mg	57	115

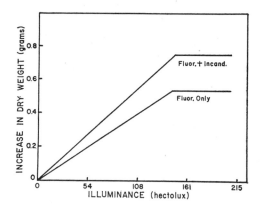

FIG. 3. Dry weight yields of tomato plants as a function of light intensity and quality (Went, 1957).

some indication, but admittedly very few data, that more than 10% incandescent illumination would be better. The lighting system of a modern plant growth chamber would be typified by the SED curve shown in Fig. 4.

The physical characteristics of fluorescent lamps create problems when they are used in plant growth chambers. The fluorescent lamp is a glass tube coated with a mixture of phosphors that fluoresce when exposed to

ultraviolet radiation. When the lamp is operated, cathodes at the ends of
the tube are heated to induce the release of electrons which ionize a filling
gas and reduce tube resistance so that an arc can be struck. The arc current
flows through mercury vapor, changing the energy levels of the electrons
in the mercury ions and releasing energy. Because the lamp is operated at
a low pressure the wavelength most efficiently generated is 254 nm and the

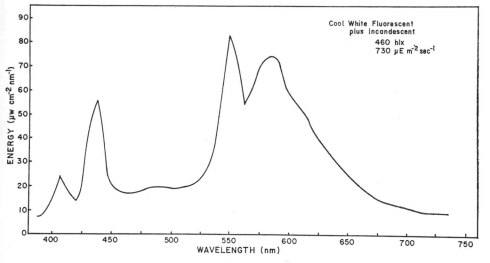

FIG. 4. Spectral irradiance 1 m from the lamps in a plant growth chamber equipped
with cool white fluorescent and incandescent lamps in a ratio of 3·8:1. Sharp peaks
indicate mercury lines.

proportion of the energy in that wavelength is very sensitive to changes in
mercury vapor pressure. The 254 nm radiation is absorbed by phosphors
carefully selected to respond most efficiently to this wavelength. More
detailed accounts of lamp operation can be obtained from Elenbaas (1959),
Allphin (1959), and IES (1968). The proportion of energy from the mercury arc in the various emission lines is determined by the pressure and
consequently the temperature at which the lamp is operated (Fig. 5). Since
more mercury is placed in the lamp than will ever be vaporized at one time,
the excess condenses at the coolest spot in the lamp. Therefore, mercury
vapor pressure, especially of highly loaded, 1500 mA lamps, is partly
controlled by lamp design. The T-12 type (Tubular, 12 eighths of an
inch diameter) has the cathodes extended into the tube so that a mercury
condensation end-chamber exists behind a shield. A lamp of non-circular
cross section, such as the Power Groove, uses two points of constricted
cross section near the center of the lamp for mercury control. Philips lamps
use a protrusion near the center of the lamp for this purpose. In every

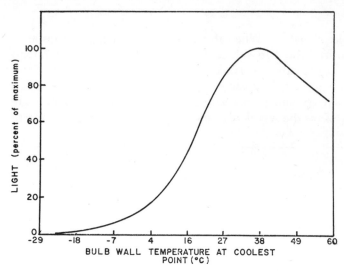

FIG. 5. Light output of a 1500 mA PG-17 fluorescent lamp as a function of minimal tube wall temperature (General Electric).

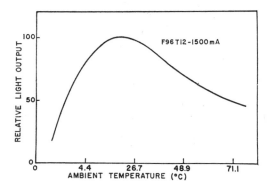

FIG. 6. Effect of ambient temperature on light output of T-12, 1500 mA lamps (GTE Sylvania Inc.).

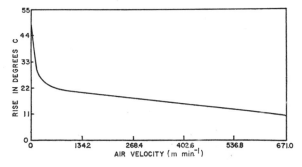

FIG. 7. Effect of air velocity on the temperature of a 1500 mA, PG-17 fluorescent lamp (General Electric).

case the design results in a part of the wall operating at a lower temperature than the rest of the tube, thereby providing a mercury condensation point.

Tube wall temperature is, of course, related to the ambient temperature and to the air flow over the tube (Figs 6, 7). In the plant growth chamber light cap the lamps are usually packed closely together, often 2–3 mm apart. Forced draft cooling therefore is necessary, although lamp design will change the susceptibility to temperature. For example, smaller diameter lamps like the T-10 have the same losses as a PG-17 but they are distributed over a smaller surface area. Consequently, the T-10 tends to run hotter and respond faster to a change in temperature. As one would suspect the optimum ambient temperature is lower for a T-10 than for a PG-17 (Fig. 8).

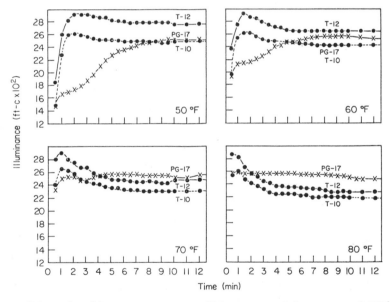

FIG. 8. Effect of ambient temperature on light output of three types of 1500 mA fluorescent lamps in a growth chamber application (unpublished data from Downs, Bailey and Klueter).

In practice the ambient temperature of the lighting system should be maintained below 25°C. If lamps become warmer or colder light output can drop 25–30% or more. The majority of plant growth chambers are constructed with the lamps in a compartment separated from the growing area by a transparent barrier. The lamp compartment is then cooled by ventilation with air from the space surrounding the CER. If that space is not air-conditioned the lamps will be too warm during the summer with a corresponding variation in light intensity. If the surrounding space is air conditioned, a ventilation system that provides 80–100 m min^{-1} of air can keep the lamps at a near optimum condition. Some chambers use a completely closed system with the air recirculating through a separate heat exchanger for temperature control. While the latter system may keep the barrier cleaner than ventilating with conditioned space air, it must be designed exactly right if serious maintenance problems are to be avoided. The direction of the air flow depends on lamp design and the inlet air should be directed to the mercury condensation point. Thus, the air should be directed to the center of PG-17 and Philips HO lamps but towards the ends of T-12 or T-10 types.

Light output from fluorescent lamps decreases with use. The most rapid rate of loss occurs during the first 100 h while the lamps are stabilizing. Since the light depreciation can be as much as 10%, measurements during this period will not be valid. Following the initial light loss, output continues to decrease throughout the life of the lamp. Very few data are available on lumen maintenance in plant growth chambers, but light depreciation seems more rapid than manufacturers' curves indicate. The decrease in light is related to phosphor efficiency. Therefore, if the lamps operate at a higher than optimum condition for any appreciable time an irreversible damage may occur that reduces output. As a result of the light depreciation the rated life of the fluorescent lamp is never reached in the plant growth chamber if proper steps are used to keep the light intensity reasonably uniform. This means a lamp changing schedule that considers allowable uniformity balanced against man-hours consumed and modified by the allowable disturbance to the experimental material. If we consider a lighting system that would produce 600–645 hlx with all new lamps we can maintain 430–500 hlx by changing the oldest one-third of the lamps whenever the illuminance reaches the lower limit. A growth chamber would then contain lamps of three ages. This system is in operation at the French phytotron (Nitsch, 1972) and at SEPEL.

New developments in the high intensity discharge lamp have prompted a good deal of interest in them as a light source for plant growth. These lamps include color-improved mercury, metal halide, and high-pressure sodium. Advantages are many in that ambient temperature does not affect light output, lumen maintenance is reported to be better than with fluorescent lamps, and automatic light level controllers are probably feasible at reasonable cost. Moreover, the area per watt of 1 compared to 16 for 1500 mA T-12 fluorescent lamps will allow a greater number of installed watts per square meter, and consequently light levels as high as 1100 hlx can be obtained. Unfortunately, very few data are available that compare the plant growth produced under various types of HID lamps with that obtained from the usual fluorescent–incandescent system. A 50:50 mixture of the light from clear mercury and the light from high pressure sodium (Lucalox) lamps was reported to produce plant growth at least equal to that obtained with fluorescent–incandescent (Table Va and b). General

TABLE Va

Effectiveness of mercury–lucalox lamps on growth of Marketer cucumber and Pinto bean

Light source	Illuminance (hlx)	Stem length (cm)		Fresh wt (g)		Dry wt (g)	
		Cucumber	Bean	Cucumber	Bean	Cucumber	Bean
Fluorescent– incandescent	430	11·0	17·8	19·9	13·0	1·70	1·05
Mercury– lucalox	215	2·0	28·6	7·0	9·0	0·52	0·71
Mercury– lucalox	430	4·6	8·8	16·6	10·5	1·30	0·82
Mercury– lucalox	860	6·6	4·2	21·2	5·8	1·70	0·36
Mercury– lucalox	750[a]	5·5	6·5	18·6	9·2	1·50	0·84

[a] With a 5 cm water filter

TABLE Vb

Effectiveness of mercury–lucalox lamps on growth of Coker 319 tobacco and Wheatland milo

Light source	Illuminance (hlx)	Fresh wt (g)		Dry wt (g)	
		Tobacco	Milo	Tobacco	Milo
Fluorescent–incandescent	430	73·4	31·2	5·54	3·69
Mercury–lucalox	215	59·8	15·6	4·03	1·76
Mercury–lucalox	430	71·9	54·0	5·34	6·44
Mercury–lucalox	860	61·3	59·0	5·32	8·17
Mercury–lucalox	750[a]	75·7	53·0	6·18	6·51

[a] With a 5 cm water filter

plant appearance was good, leaf size satisfactory and internodes adequate, but not elongated. In fact, the mercury–Lucalox combination produced nearly twice the weight per unit length as did the fluorescent–incandescent system. Either lamp alone, however, resulted in plants that were unsatisfactory at least in appearance. Comments appearing in the literature that

Lucalox lamps appear to be the most efficient are extremely misleading. They undoubtedly are the most efficient producers of light for vision and perhaps of PAR. This does not mean that spectral distribution will properly satisfy all the photoreactions that regulate plant growth, and consequently plant appearance may be unsatisfactory. Of all the HID lamps the metal halide ones are the most likely to produce satisfactory plant growth. A number of such lighting systems are in operation, although few comparative data on lamp or plant behavior seem to be available (Bretschneider-Herrmann, 1962, 1969; Chandler, 1972; Kawarada and Shibata, 1972; Mitchell, 1972).

SPECIALIZED LIGHTING

Although the biologist has an interest in lasers, polarized light, point sources etc., the main emphasis has been on methods of obtaining monochromatic light. Since the use of monochromatic light continues unabated some discussion of the subject seems appropriate.

Monochromatic light is supposed to be composed of only a single wavelength but in practice a more or less narrow band of wavelengths is obtained. Narrower regions of the spectrum can be produced by spectrographs or by monochromators. Broader spectral regions are produced by combinations of filters and light sources.

Spectrographs. A spectrograph uses a high intensity source such as carbon or Xenon arc lamps in conjunction with prisms or diffraction gratings to produce a spectrum. An optical system of lenses, mirrors, and slits is used to collimate the light beams. The large spectrograph once used at Beltsville (Parker *et al.*, 1946) is an excellent example of a prism type instrument, and grating types are in use at the Argonne Lab (Monk and Ehret, 1956), the Biotron Institute in Japan (Matsui *et al.*, 1971), CERES in Australia (Annual Report 1972), and the phytotron in France (Nitsch, 1972). Small spectrographs such as the one described by Norris (1968) can be built with wedge-interference filters. In each type of spectrograph the spectrum produced is continuous and is used to irradiate biological material with energy simultaneously in many spectral regions.

MONOCHROMATORS

In practice monochromators usually produce a narrow band of wavelengths rather than, as the name implies, a single one. The most elaborate monochromators use the grating or prism spectrograph to project a select portion of the spectrum onto one or more exit slits which determine the wave band of the resultant monochromatic light (Balegh and Biddulph, 1970). Other types use narrow-band pass filters with an appropriate light source and optical system (Withrow, 1957; Mohr and Schoser, 1959). In most biological studies more than one wavelength region is investigated so that several monochromators would be required.

Wide band filters. The most common form of monochromator has filters that pass a rather broad segment of the spectrum. Gelatin filter (Withrow and Price, 1953) monochromators and the red–blue cellophane filters (Downs *et al.*, 1958) used in early phytochrome studies are excellent examples. The wide band filter provides a distinct advantage to those investigating whole-plant physiology by allowing irradiation of large areas with relatively high energy levels.

Practical wide band monochromators are easy to construct. For example, a very useful red light source of good purity can be obtained with 2444 red plexiglass in a commercial fluorescent lighting fixture (Fig. 9). The filter

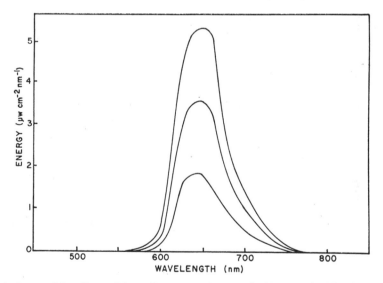

FIG. 9. Spectral irradiance 75 cm from two, four or six 40 W cool white fluorescent lamps filtered by red plexiglass.

transmits the small amount of far-red emitted by the fluorescent light but the effect is usually negligible in the presence of the much higher level of red. However, blue or green plastics, which also transmit far-red, must be used in conjunction with a liquid filter containing copper sulfate because the low level of far-red can have a measurable biological effect.

Transmittances of water, copper sulfate and ferrous ammonium sulfate solutions were published by Withrow and Price (1953) and these liquid filters can be used alone or in conjunction with wide band filters.

Far-red systems use a special plexiglass called FRF-700 developed by Rohm and Haas Co. in cooperation with the Beltsville groups (Downs et al., 1964). A simple but effective far-red luminaire shown in Fig. 10 was des-

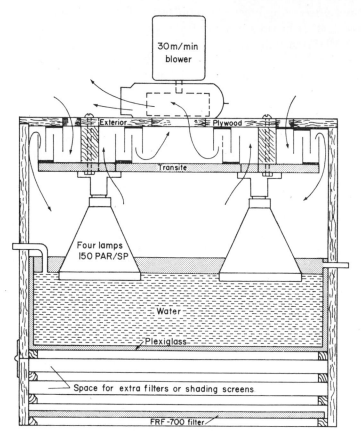

FIG. 10. Far-red luminaire designed for studies of phytochrome physiology. The system provides an irradiance at 740 nm of 24 μW cm² at a distance of 78 cm (Downs and Bailey, 1967).

igned by W. A. Bailey. Since the plexiglass filter does not remove long wavelengths the water bath prevents radiant heating of the biological material during irradiation.

There are many applications for narrow spectral regions where a monochromator cannot be used. In these cases efforts have been made to use special light sources. Most single-color fluorescent lamps have tended to fade and failed to maintain the original spectral distribution. However, excellent red lamps have been produced using magnesium germinate or arsenate phosphors and recently Westinghouse has made an improved blue lamp. Far-red can be obtained from incandescent lamps with a natural dark ruby envelope. The largest of these, the BCJ, is only 60 W and although the emission extends rather far into the red region (Fig. 11), they have many

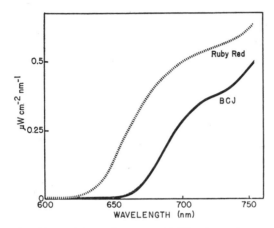

Fig. 11. Spectral irradiance 75 cm from one 60 W ruby red or one 60 W BCJ incandescent filament lamp (Downs *et al.*, 1964).

applications in research. Wide band filters and special light sources are often used directly in or as part of the plant growth chamber. Vince and Stoughton (1957). Meijer (1959), and De Lint (1960), for example, used such systems to study spectral dependence of flowering and morphogenesis.

Photosynthesis

The biochemical and biophysical details of photosynthesis have been admirably discussed by a number of authors (Kamen, 1963; Clayton, 1965; San Pietro *et al.*, 1967, Rabinowitch and Govindjee, 1969; Hatch *et al.*, 1971; Zelitch, 1971). While these details are of utmost importance to our understanding of the photosynthetic mechanism, we are mainly concerned here with how light controls the end result—productivity in terms of plant growth. Thus, the term photosynthesis is used here in a broad sense to mean the way in which light induces an increase in growth usually measured in dry weight.

Sunlight produces a maximum illumination of 1000–1300 hlx on clear days when the solar angle is near one. Using data from various sources (van Wijk, 1963; Army and Greer, 1967; Bjorkman and Ludlow, 1972) the maximum energy can be estimated to reach $1\cdot45$ cal cm^{-2} min^{-1} which would provide about $0\cdot64$ cal cm^{-2} min^{-1} of photosynthetically active radiation (PAR). These values represent peak energies so they are available for only a small portion of the light period on only a certain percentage of days. Nevertheless, many biologists believe energies of this magnitude are essential in the plant growth chamber. Certainly the ability to achieve solar energy levels may be advantageous for some kinds of research. However, the cost of obtaining and using such high energies would seem to make them impractical for general applications of controlled-environment rooms; especially if the natural energy relationship have been misinterpreted and if the majority of plant species reach a saturation point at much lower light levels.

The range of maximum energy values is certainly of ecological interest and Gaastra (1964) suggests that frequency of distribution is even more important than range. If the maximum value only occurs on a few percent of days then it really is not a very reliable indicator of energy received. As expected, few growth chamber lighting systems reach the sunshine levels obtained on even 70% of the days during the growing season (Table VI). Gaastra's (1964) data indicate, in fact, that controlled environment

TABLE VI

Radiant energy between 400–700nm in Wageningen compared to flux densities in growth rooms (adapted from Gaastra, 1964).

| Solar energy (cal cm^{-2} day^{-1}) | | | Energy at noon (cal cm^{-2} min^{-2}) | |
Month	Range	On 70% of days	Range	On 70% of days
April	20–220	160	0·03–0·52	0·36
June	50–280	220	0·04–0·60	0·43
July	60–260	200	—	—
August	50–230	170	0·04–0·57	0·37
September	30–170	130	0·04–0·49	0·30
December	5–40	25	0·01–0·14	0·07
HPL—400[a] lamps (3200 W)		275/16 h day		0·27
TL —120[b] lamps (1780 W)		185/16 h day		0·19

[a] Philips high pressure mercury lamps with fluorescent coating
[b] 120 W white fluorescent lamps

room lighting would equal natural light on only 40–45% of the days.

The total energy received by the plant each day presents a somewhat different picture of light relations. Approximately 500 cal cm^{-2} are received on a clear day during the growing season. This results in about 220 cal cm^{-2} day^{-1} of PAR with an estimated quantum flux of 4320 microeinsteins

cm^{-2} (Army and Greer, 1967). The values vary with latitude and atmospheric conditions, i.e. Bjorkman and Ludlow (1972) report a daily PAR of 5890 microcinstcins cm^{-2} in Queensland, and serve here only as illustrations. The day-to-day variation is, of course, considerable but unlike the maximum values the total energy on 70% of the days can be equalled or exceeded in the plant growth chamber.

The optimum amount of light for plant growth should be attainable from data relating to saturation values for net assimilation. The literature on this subject begins at least as early as 1883 and the reported values range from 100 to 1000 hlx. Went (1957) shows that the dry weight production of tomato plants failed to increase after the illumination exceeded 137 hlx (Fig. 3). Bohning and Burnside (1956) indicate that 215 hlx would allow the maximum rate of photosynthesis in sun plants. Blackman and Black (1959) divided plants into shade, intermediate and sun types with maximum photosynthetic rates at 108, 108–215, and 125–323 hlx respectively. Sweet and Wareing (1966) found that saturation for *Pinus radiata* was essentially at 322 hlx. Some species however seem to require as high as 645 hlx for saturation (Blackman and Black, 1959) and Klueter *et al.* (1971) state that photosynthesis of cucumber plants had not leveled off at 860 hlx.

There is no doubt that measurements of saturating levels of light for photosynthesis are factual. Nevertheless, such data are often confusing and misleading without considerable detail of experimental methods. Many of the data refer to CO_2 utilization of a single leaf at right angles to the light source. Measurements are often of brief duration and may have included a starvation period of several hours in darkness to deplete the carbohydrate in the leaf. Considerable caution is called for when the results of such experiments are extrapolated to whole plant behavior.

The net assimilation of a canopy of plants is the sum of the photosynthetic activity of each leaf and that activity is a function of age and illumination. Photosynthetic efficiency decreases markedly with the age of the leaf. Burton (1972) showed that potato leaves that are three-quarters yellow-green have less than half the rate of CO_2 utilization when compared with recently expanded leaves. In Phytotron studies with tobacco plants CO_2 absorption began to decrease soon after topping. Although the upper leaves were continuing to expand the lower ones were becoming senescent so that the total photosynthetic efficiency was less.

Self-shading also complicates the light saturation information when whole plants are considered. Went (1957) for example, noted that while only 108 hlx was required to saturate a single plant, a population required 473 hlx for maximum photosynethesis. Similarly Hoover *et al.* (1933) and Friend *et al.* (1963) showed that 215–270 hlx are required for maximum assimilation of a single wheat plant whereas in the field the light requirement rises to 807 hlx (Thomas and Hill, 1937). Leafe (1972), studying a

50×50 cm plot of rye grass, found that the saturation irradiance was 0·6 cal cm^{-2} min^{-1} 10 days after clipping and increased to 0·9 cal cm^{-2} min^{-1} after 35 days. The increase was largely due to shading and leaf age.

Saturation light levels for photosynthesis may change for a given plant species grown under different light levels prior to measurement. Bjorkman *et al.* (1972) illustrated this point with *Atriplex* where the photosynthetic characteristics adjusted to the radiant flux density during growth. Thus, the irradiance required for saturation of plants grown at high light levels was several times greater than for plants grown at lower ones. Similar responses have been noted for pasture plants such as *Panicum maximum* and *Phaseolus atropurpurea* (Ludlow and Wilson, 1971) and Burnside and Bohning (1957) reported a marked decrease in saturation irradiance for several crop plants when they were grown at reduced light intensities.

Short term measurements may also be misleading since CO_2 utilization and/or dry weight production may not be constant throughout the growing period even under constant light conditions. Moreover, a measured rate of photosynthesis may not result in a corresponding increase in dry weight production. Warren Wilson (1972) showed that a sevenfold increase in light intensity caused a threefold increase in the net assimilation rate which remained at the new, high rate for the next 10 days. Growth also increased greatly the first day after transfer to the higher light intensity but then decreased each day thereafter until it reached a new equilibrium that was only slightly greater than the original rate at the lower illuminance. Very little of the increased growth showed up as structural weight. The increase was mainly in the storage fraction of the dry weight; material removed by hydrolysis in hot sulfuric acid.

Dry weight production or growth is a function of the duration of the light as well as the radiant flux density. However, total energy does not provide an adequate description for the growth produced since the reciprocity law, $E = It$, does not hold. For example, Went (1957) states that at 108 hlx dry weight production will double when the duration of the light period is increased from 8 to 16 h but not when the intensity is doubled to 215 hlx. Other data (Fig. 12, Table VII) show quite clearly that a lower intensity for a long time is somewhat more effective than double the intensity for half the time. If the time at the higher 430–538 hlx light level is doubled even more growth is obtained but the plants may have an abnormal appearance because other factors such as nutrition and CO_2 level may begin to interact adversely.

In the plant growth chamber plants are usually spaced to minimize overlap and the large amount of side light substantially reduces self-shading effects. Therefore, plants in a modern plant growth room with a light system that produces 430–538 hlx rarely will have leaves at less than 215 hlx. Thus, for many kinds of plants the light saturation level for

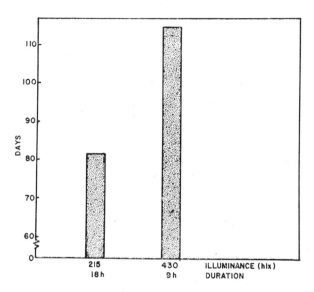

Fig. 12. Effect of light intensity and duration on time of development of the first inflorescence in Carefree White geraniums (adapted from Biamonte, 1972).

TABLE VIIa

Growth of three varieties of tomato under three different light regimes

Variety	Fresh weight (g)		
	484 hlx; 16 h	484 hlx; 9 h	247 hlx; 16 h
Manapal	31·8	21·5	22·2
Fireball	37·8	27·4	33·5
Small Fry	39·6	26·3	30·9

TABLE VIIb

Growth of three marigold varieties after 4 weeks at the same total energy given in two sequences

Variety	9 h at 430 hlx		18 h at 215 hlx	
	Fresh wt	Dry wt	Fresh wt	Dry wt
	g	mg	g	mg
Sparky	5·8	360	8·8	610
King Tut	5·4	390	8·9	600
Spanish Brocade	6·8	460	9·1	630

dry weight production is approached. Data using Hg-Lucalox lamps indicate that the saturation plateau is essentially reached between 215 and 430 hlx with the plants in growth room-like conditions and spacing (Jividen et al., 1970).

Irradiances currently available with fluorescent-incandescent sources are adequate for most plant growth purposes. Where additional energy is necessary the HID lamps can be used although the additional energy will increase initial as well as operating costs. Moreover, at high light levels radiant heat can become a serious problem and a water bath with all its attendant maintenance problems may be a necessity. Biamonte (1972) (Table VIII) shows the effect of illuminance on geranium seedlings. The 700 hlx illuminance was under a water bath. Without the water bath the seedlings died. Plant temperature increases (Table IX) do not appear to

TABLE VIII

Effects of light intensity on time of inflorescence development and heights of Carefree White geraniums (Biamonte, 1972)

Illuminance (hlx)	Stage 1[a] days	height (cm)	Stage 2 days	height (cm)	Stage 3 days	Inflorescence number
215	94	40	121	50	133	2·6
430	40	13	63	25	79	3·8
700	46	16	70	27	89	4·5
LSD$_{05}$	15	7	15	7	15	1·1

[a] Stages of inflorescence development: (1) macroscopically visible; (2) first open floret; (3) all florets open

TABLE IX

Temperatures of light and dark periods in the growth chamber cubicle of 800 hlx and 700 hlx recorded in the soil, the air at the upper leaf canopy, and underneath the uppermost leaf

	Temperatures			
	800 hlx Light °C	Dark	700 hlx Light °C	Dark
Soil	33	20	27	20
Air	31	20	27	20
Leaf	29	20	25	20

be the entire explanation because geranium seedlings do not die at 31°C air temperature and a lower light level.

EFFECT OF LIGHT QUALITY ON PLANT GROWTH

Of all the light sources available, the high pressure sodium lamp is undoubtedly the most efficient producer of light. McCree (1972) indicates that such sodium lamps also have the highest photosynthetic efficiency;

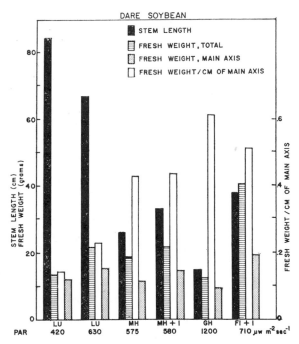

FIG. 13. Growth of Dare soybean after 35 days under sodium (Lu) metal halide (mh) or metal halide plus incandescent lamps as compared to that obtained in a fluorescent–incandescent growth room or in the greenhouse (day/night temperature 26/22 °C).

roughly 8 % greater than the sun per unit of quantum flux density between 400–700 nm. Most other commonly used light sources such as the fluorescent or the metal halide and mercury HID lamps have about the same photosynthetic efficiency as the sun. However, these lamps have very different spectral energy distributions and from the work of a number of investigators (Wassink and Stolwijk, 1956; Meijer, 1959; De Lint, 1960) one could assume that plants would not grow equally well under these light sources. Generally plants grown under light sources that produce very little blue are elongated, often with small leaves, and appear much like plants grown in the shade. Plants grown under light sources largely devoid of red appear normal except for considerably less height growth. Figure 13 compares the growth of soybean plants under various light sources. Fresh weight was comparable between plants grown under sodium and metal halide plus incandescent lamps. However, in view of excessive elongation under the sodium lamps, the sturdier (greater fresh weight per unit length) plants from the metal halide and incandescent sources were much more desirable.

Regulatory Light

Success in growing plants in controlled-environment rooms is enhanced by and often depends upon knowledge of the role of light in regulating plant growth. The regulatory effects of light are of course discussed in plant physiology and other plant science courses but it is our experience that the entire subject is often received as interesting but purely academic. Stress should be laid on the fact that photobiological regulation is a very practical system which operates in nature and in controlled-environment facilities to control plant growth and development.

To understand the practical application of growth regulation by light it is necessary to have some knowledge of the pigment systems involved and how they work. Since our discussion of regulatory light begins with germination, it is here that we shall explain how the photoregulatory system functions. This seems appropriate because a great deal of our knowledge of the photoreactions controlling plant growth has been obtained through seed germination studies and many of the photobiological actions discovered during germination also occur later during morphogenesis and flowering.

GERMINATION

The germination process begins with the imbibition of water. Enzymes become activated and an increase in respiration is followed by cell elongation and/or cell division. If one or more of the processes leading to germination is blocked and does not proceed, the seed fails to germinate and is sometimes referred to as dormant. Temperature, chemicals and light may block one or more pathways to cause dormancy and the same factors may remove such a block. The light reaction becomes operative as soon as the water content of the seed begins to increase; in some cases at moisture levels as low as 15% (Hsiao and Vidauer, 1971).

Light-requiring seeds are not at all unusual, although fewer light-sensitive seeds are found in cultivated species than in wild ones. Yet the Biological Handbook (1962) notes that of eighty-nine cultivated herbaceous species light is either required or substantially enhances germination in thirty-five. A sample of the Bromeliaceae that included twenty-three species representing ten genera showed light to be required by seventeen species, and enhanced the germination of three others. Only the three species of *Tillandsia* showed no response to light (Table X).

TABLE X

Germination of seeds of several species of Bromeliaceae under several light conditions at 23°C (Downs and Piringer, 1958)

Species		Period of germination	Darkness	Seeds germinating when exposed to:			
				8 h of light daily	1 h of light daily	20 min of light every 8 h	5 min of light every 2 h
Pitcairnia sp.:	Expt. 1	18	0	0	0	4	88
	Expt. 2	26	0	12	0	6	98
	Expt. 3	28	0	18	0	0	91
	Expt. 4	34	0	84	83	—	—
Tillandsia pulchella		30	80	100	—	—	—
T. geminiflora		30	64	66	—	—	—
T. stricta:	Expt. 1	30	80	78	—	—	76
	Expt. 2	30	100	100	96	93	96
Vriesia regina		26	0	100	30	92	85
V. incurvata		28	26	64	—	—	70
V. geniculata		22	0	90	17	83	48
V. lubbersii		22	40	60	60	48	60
V. philippocoburgii		18	0	97	100	100	69
V. philippocoburgii var. *vagans*		18	0	51	36	32	66
V. haematina		18	0	96	96	100	92
V. petropolitana		18	43	97	93	100	91
V. ensiformis		18	0	63	59	63	75
Neoregelia indecora		7	0	—	—	—	93
N. concentrica		7	0	98	63	—	96
N. carolinae		11	0	88	28	—	88
Nidularium fulgens		11	0	98	98	—	98
Aechmea coelestis		10	0	83	65	95	91
Canistrum lindenii var. *roseum* forma *procerum*		11	0	88	80	—	98
Billbergia sanderiana		7	0	96	96	96	96
B. elegans		7	0	100	100	—	100
B. pyramidalis		7	0	100	100	—	—
Quesnelia lateralis		7	0	98	98	96	100

Seeds that are not light requiring may become so under a number of conditions. Nutile (1945) induced a light sensitivity in non-light-requiring lettuce seeds by the action of coumarin. Lettuce seeds also become light requiring if they are held for a few days at 30–35°C in darkness. At 35°C germination is prevented but when the temperature is lowered the germination processes resume. When the high temperature period is prolonged, apparently an additional block develops so that when the temperature is lowered germination will not occur without an exposure to white or red light (Toole *et al.*, 1953).

Non-light-requiring seed of Great Lakes lettuce, tomato, *Lamium amplexicaule, Nemophila insignis* and others can be prevented from germinating by a long exposure to continuous far-red. After inhibition by far-red radiation many of the seeds become light requiring (Hendricks *et al.*, 1959; Mancinelli *et al.*, 1967).

Light sensitivity can also be induced in seeds by holding them on a medium with a high osmotic concentration (Kahn, 1960) for more than 30 h. When transferred to water under an appropriate safe light the seeds still fail to germinate but they now respond to light in the same manner as seeds made light dormant by temperature or by far-red radiation.

Many kinds of seeds respond to only one brief light exposure but many others require more than a single irradiation, and in many cases require relatively long periods (Isikawa, 1954; Black and Wareing, 1955). A number of species in the Bromeliaceae, for example, failed to germinate completely with 1 h exposures per day, whereas 8 h of light resulted in maximum germination. Since so many kinds of seeds are induced to germinate with brief irradiation periods it seems unlikely that such long exposures were necessary. A detailed study with seeds of *Puya berteroniana* showed that two 15 min light periods separated by 12 h of darkness induced an amount of germination equal to that obtained with 8 h of light per day. A 30 min irradiation once a day was largely ineffective. Reducing the duration of both exposures below 15 min resulted in a decrease in germination. However, one of the two daily exposures could be reduced to 4 min without losing germination capacity as long as the other exposure remained 15 min (Table XI). Many other kinds of seeds, such as additional Bromeliad species (Table X) and *Paulownia tomentosa* (Borthwick *et al.*, 1964) that appear to have long irradiation requirements also respond to cyclic lighting.

The reaction of seeds to light is controlled by phytochrome, a ubiquitous pigment found in all higher plants and in ferns, mosses, and algae. The phytochrome molecule is a biliprotein with two forms, inconvertible by light. The chromophore seems to be a bilitriene similar to that of *c*-phycocyanin and allophycocyanin (Siegelman *et al.*, 1966). The red form, P_r, absorbs light at about 660 nm and is converted to the far-red form, P_{fr}, which has an absorption maximum at about 735 nm. A number of

TABLE XI

Effect of light and temperature on the germination of
seeds of *Puya berteroniana*

| Light | | % of seeds germinating at indicated temperatures (°C) | | | |
8 a.m. min	4 p.m. min	16	20	22	25
0	0	0	0	0	0
19	0	21	4	0	0
4	15	92	72	12	8

red intermediate forms have been demonstrated and far-red intermediates probably occur.

The P_{fr} form is usually considered the biologically active form. It is converted to the P_r form slowly in darkness. The absorption of the P_r and P_{fr} forms overlap so that a photostationary state is induced. Since the extinction coefficients are high the photostationary state is established very rapidly. The P_r–P_{fr} ratio and consequently the percent P_{fr} present will be a function of the R–FR ratio in the light source (Table XII). Differences

TABLE XII

Light source	R–FR Ratio	% PFR
Red Bank	16·5	76–78%[a]
Fluorescent, cool white	7·4 (5·0[b])	
Incandescent, white frosted	0·68 (0·74[b])	60[c]
Incandescent, ruby red	0·489[c]	40[c]
Incandescent, BCJ	0·189[c]	21[c]
Far Red Bank	0·027[c]	1·2–2·5[a]
Cool white Fluorescent and Inc.	2·0	

[a] extracted phytochrome in solution (Mancinelli and Downs 1967)
[b] Downs *et al.*, 1964
[c] Evans *et al.*, 1965

in the data can be a result of method or instrumentation. For example, the R–FR ratio from fluorescent lamps is 7·4 with a narrow band pass instrument and 5 with one that uses a greater band width. A ratio of 0·74 on a nanometer basis may become 1·07 if a larger absorption area of each phytochrome form is used. Whatever the measurement difficulties a steady state level of P_{fr} is determined by the light source and is maintained as long as the light is on. When the light is turned off, P_{fr} continues to act but the level decreases as it reverts to the P_r form. It is possible that dark reversion occurs only after reaction with a substrate. The percent conversion of P_r

to P_{fr} required to produce a given biological response differs with the various responses and with plant type. For example, lettuce seeds need a 50% conversion for 50% germination, whereas *Lepidium* only requires 17% conversion. The incident energy necessary for a particular fractional conversion seems to vary with conditions although the corresponding germination is constant.

In some seeds P_{fr} needs to be produced only once and the dark conversion rate is so slow that P_{fr} is present above the critical level for sufficient time to start the germination processes. Seeds such as *Puya* and *Paulownia*, however, require two or more conversions per day, usually for several consecutive days, in order to provide continuous P_{fr} action for the required time. If the P_{fr} is removed by a far-red irradiation before it has had sufficient time to function the promotive effect of the red radiation is nullified.

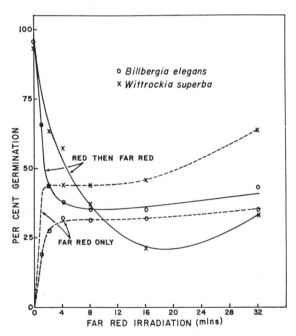

FIG. 14. Effect of far-red or far-red following an exposure to red on germination of seeds of *Billbergia elegans* and *Wittrockia superba* (Downs, 1964).

While non-light-sensitive seed such as Great Lakes lettuce and tomato are inhibited from germination by continuous far-red radiation, and in fact become light requiring as a result, brief exposures to far-red usually have little effect. Some types of light-sensitive seed however are promoted by brief exposures to far-red although not to the same extent as with red (Fig. 14). These kinds of seeds can often be identified by the failure of far-red radiation to reverse completely the promotive effect of an exposure to red light. Absorption characteristics of phytochrome (Table XII) indicate that far-red radiation would result in a low level of P_{fr} which must be enough to promote germination in some seeds of the population. Additional seeds will germinate as the low level of P_{fr} is maintained by continuing the irradiation; this could indicate a substrate limitation for the P_{fr} in some of the seeds since the level of P_{fr} is proved adequate by the initial germination.

Knowledge of the light requirements of seeds and the details of the response can assist in better plant production through increased amount and uniformity of germination. Moreover, light can be used as a tool to set and control the physiological conditions of seeds for other types of studies. Light requirements for germination also play an important, although little investigated, role in natural and crop ecology. The different spectral distribution under a canopy of plants (Table XIII), for example, could

TABLE XIII

Ratio of red and far-red irradiances under two kinds of plant canopies

	660/740 ratio
Open: full sun	1·33
Virginia pine stand	0·50
Mixed hardwood stand	0·35

induce a quite different germination response than would be obtained in the open. Tests using *Catalpa* leaves as filters resulted in a definite far-red effect which could inhibit germination and make a non-light-sensitive seed light requiring if it was under such a leaf canopy. Certainly many observations show that non-light-sensitive seed do become light requiring in nature. When some kinds of seeds are freshly harvested they germinate equally well in light or darkness. However, under natural conditions when weed seedlings are pulled out of the plant bed or the soil surface disturbed by other means, they are shortly replaced with new ones which come from seed lying dormant in the soil, and which one might assume were activated upon exposure to light resulting from disturbing the soil surface. If the seedlings are killed without disturbing the soil few seeds germinate. Thus, while it may be possible that the seedlings themselves in some way inhibit germination and even create conditions that induce the light requirement, the release of that inhibition by killing or removing the seedling is not in itself sufficient to allow germination.

A knowledge of how light operates to control seed germination helps us to understand how phytochrome works and enables us to use the system more effectively in regulating other plant responses.

PHOTOPERIODISM

Flowering. Natural variation in the relative durations of the day and night periods (Fig. 15) is a major factor in the seasonal flowering of many kinds of

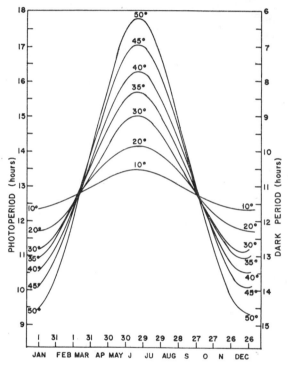

FIG. 15. Annual change in light and dark periods at 0–50° latitudes (American Nautical Almanac, 1943).

plants. Although it may seem rather obvious today that daylength would exert an influence on plants and animals, it was not until 1920 that the phenomenon was clearly pointed out (Garner and Allard, 1920). They classified plants as short day or long day depending upon their response to photoperiod. Plants that did not respond to photoperiod were called day-neutral and a few species were classed as intermediate, flowering only on a specific range of daylengths. By definition a short day plant is one that flowers on any daylength less than a certain critical number of hours, and a long day plant is one that flowers on photoperiods greater than the critical one. Many plants, however, do not remain vegetative on adverse daylengths but will eventually flower. These are considered facultative photoperiod-sensitive plants. Some like poppy (*Papaver somniferum*) are obligatory long day plants when small but as they increase in size become facultative and eventually flower on short days. Flowering of *Celosia argentea* is accelerated by short days but it seems to flower eventually on long days. A close inspection of the plumose type collected in Brazil reveals that on long days very few flowers and consequently few seeds are actually produced. The apparent inflorescence is composed almost entirely of colorful bracts (Table XIV).

TABLE XIV

Influence of daylength and temperature on the production of fertile flowers by *Celosia argentea*

Temperature °C Day/night	Daylength (h)	No. of inflorescence branches	No. of flowers Main Axis	Branches
26/22	9	37	16	220
26/22	9 + 3[a]	78	12	16
22/18	9	84	9	106
22/18	9 + 3[a]	70	0	3

[a] 3 h interruption near the middle of the 15 h dark period

Commercial cultivars of the head type Celosia like the Toreador variety also respond well to short days. When these plants eventually flower on long days it may be accompanied by severe stem fasciation. Fasciation of strawberry also occurs on certain photoperiods (Table XV).

TABLE XV

Effect of photoperiod on reproduction of two strawberry varieties (Downs and Piringer, 1955)

Photoperiod (h)	Flower clusters No. Climax	Red Rich	Runners No. Climax	Red Rich	% of plants fasciated Climax	Red Rich
11	5	5	16	0·5	28	100
13	3	5	33	1·7	0	85
15	0	14	55	1·6	0	0
17	0	23	43	1·4	0	0

Plants do not become photoperiodically sensitive at the same stage of development. Some, like *Chenopodium rubrum*, will produce visible flowers 6 days after beginning imbibition when the seeds are germinated on 8 h photoperiods (Cumming, 1959). *Pharbitis nil*, a species used a great deal in

photoperiod research, has receptive cotyledons and is responsive to photoperiod as soon as these expand—about 4 days after germination (Nakayama, 1958). Barley grown on long days from emergence will complete the
differentiation of all nodes in about 17 days. The first spikelet primordia
can be detected at the seventh or eighth node. Since there are 3 or 4 nodes
already differentiated in the seed, the photoperiodic effect must have
occurred very early (Borthwick *et al.*, 1941). In most plants the cotyledons
are rather insensitive and the leaf is generally accepted as the organ of
perception. With Biloxi soybean, for example, maximum sensitivity to
daylength seems to be at the third compound leaf (Borthwick and Parker,
1940).

The rapid rate of development begun by the inductive daylength may
persist as a recognizable after-effect when the plants are returned to non-
inductive conditions. Cocklebur (*Xanthium pensylvanicum*) plants, for
example, need only a single inductive short day to initiate flower primordia.
Moreover, the primordia proceed to develop to anthesis on non-inductive
photoperiods. Soybeans, however, which require several short days for
initiation and development, may fail to proceed if the plants are returned to
long days without a sufficient number of short ones. *Caryopteris incana* is
day neutral for flower initiation and development to bud but further progress to anthesis requires the days to be short (Piringer *et al.*, 1963). If
kept in the bud stage too long the ability to develop further is lost. Thus,
only the buds on about the upper 5 nodes usually proceed to anthesis when
placed on short day conditions. Flowering *Coffea arabica* is somewhat the
reverse of *Caryopteris*. With coffee plants, short days are required for
initiation of flower primordia but development stops at the bud stage. Bud
to anthesis development is not photoperiodically sensitive but develops in
response to water relations. Flower buds can be induced to bloom by
alternate periods of water stress and copious watering. Thus, anthesis would
occur within about 2 weeks after the plants were placed under a water
stress for several days and then watered heavily (Piringer and Borthwick,
1955). Flowering of snapdragon (*Antirrhinum majus*) is reported to be
sensitive to daylength only during a definite period of 40–64 days after
germination when the plant has 5–10 pairs of leaves (Langhans and Maginnes, 1962).

The photoperiod response is progressive, not an all or none affair as
might be concluded from results with plants like cocklebur (*X. pensylvanicum*), poppy (*Papaver somniferum*), and ryegrass (*Lolium temulentum*) that
require only a single inductive photoperiodic cycle to induce flowering.
Early studies of photoperiodism which presented data as plus for reproductive and minus for vegetative also pictured daylength control as a go–no go
phenomenon. Inductive cycles are accumulative; thus, while one short day
will induce flower initiation in some plants, three short days will result in a

much more rapid development of the inflorescence. The greater the stimulus the larger the response so the further the inductive cycle gets from the critical photoperiod the greater the inflorescence development. Thus, short day Lee soybean goes from planting to maturity in 170 days on fourteen $\frac{1}{2}$ h photoperiods but in 95 days on thirteen 1 h ones (Johnson *et al.*, 1960). Planting to awn emergence for Wintex barley needs 78, 37, and 34 days on 12, 16, and 20 h photoperiods respectively (Borthwick *et al.*, 1941) and *Petunia* will produce eight flowers per plant on 16 h days while only four are forming on photoperiods of 12 h (Piringer and Cathey, 1960).

The accumulative effect and the degree of stimulus can, however, be overdone. Cocklebur (*X. pensylvanicum*) grown on 8 h days flowers and produces burs very quickly. However, the plants are small and only a few abnormally small burs are produced. If the daylength is 13–14 h flowering will be delayed but yield and bur size increase dramatically.

Although flowering is the plant response usually associated with photoperiodism other aspects of plant behavior can also be controlled by daylength. Runner production in strawberry (Table XV), bulbing of onions and tuberization of potato and *Begonia* are examples. *Pitcairnia heterophylla* forms an above-ground bulb with modified leaves on short days. The formation of this bulbous structure is a prerequisite to flowering but flowering itself is apparently independent of daylength control. Procumbent forms of chickweed will develop on some photoperiods and cleistogamous flowers may form on certain intermediate daylengths in plants such as chickweed and soybean.

Dormancy. Cessation of shoot growth of many woody plants in the early autumn is brought about by the naturally decreasing daylengths. In the greenhouse in the fall and winter or in controlled-environment rooms on daylengths of 12 h or less, trees and shrubs will usually prepare for the onset of winter by stopping shoot growth, setting buds and hardening off (Table XVI). If the temperatures are above 20°C, leaves of deciduous species will generally remain on the "dormant" plants for a long time after growth stops.

The term dormancy is used here simply to indicate a cessation of shoot growth. *Weigelia florida* is put into a resting condition by short days which can be alleviated anytime the photoperiod is increased to 16 h or more. *Catalpa bignonioides* is first put into a rest condition by the short days but about 30 days after shoot growth has stopped a transfer to long photoperiods has no effect. Over this period of time there is a slow build-up of a dormant condition that requires cold for release. Actually, several days at temperatures of 35–40°C will also release the *Catalpa* type of dormancy and in practice the high temperatures may often be obtained more conveniently than low ones.

TABLE XVI

Effect of daylength on the growth of woody plants

Species	Stem length (cm)			
	8 h	12 h	14 h	16 h
Viburnum carlesii	34	45	85	117
Weigelia florida	2	7	25	48
Acer rubrum	4	10	34	37
Catalpa bignonioides	2	12	32	35
C. speciosa	1	2	9	23
Liriodendron tulipifera	1	2	4	21
Paulownia tomentosa	5	7	12	17
Betula mandshurica	20	38	85	140
Liquidambar styraciflua	1	3	12	18
Pinus sylvestris	2	16	37	38
P. virginiana	7	12	23	49
Pseudotsuga menziesii[a]	—	12	29	140
Picea glauca[a]	—	8	42	59
P. sitchensis[a]	—	4	5	9
Larix leptolepis[a]	—	21	27	117
Sequoia sempervirens	—	60	73	78

[a] includes branches

The first noticeable aspect of the effect of short days is that the leaves of most broad-leaved species become darker in color and the fascicular leaves of pine seedlings stop elongating. As the elongation of the stem becomes affected by daylength, internodes become shorter and shorter and finally form either a bud with true bud scales (e.g. *Acer rubrum*), or a pseudobud (e.g. in *Liriodendron tulipifera*, where the stipules function as bud scales, or in *Catalpa bignonioides* where the terminal meristem is abscised) (Downs and Borthwick, 1956). Growth of woody plants from the tropics such as *Rauvolfia vomitoria* (Piringer *et al.*, 1958) *Coffea arabica* (Piringer and Borthwick, 1955) and *Theobroma cacao* (Piringer and Downs, 1960) are also subject to daylength control and in some cases, as with *C. arabica*, flower initiation is also a photoperiodic response. These tropical plants, however, do not stop growth and enter some kind of dormant condition on short days. The growth rate is simply slowed by short days and accelerated by long ones.

Many woody plants can be kept growing continuously by long days although the critical photoperiod may vary considerably. When the days are long enough for growth to occur, the rate usually increases with increasing photoperiod. However, certain intermediate daylengths can alter the pattern of growth. For example, plants such as pine, that have a flushing habit of growth, lose that characteristic and grow continuously on

14 h days. Apparently daylength control can adjust and match rates of differentiation to rates of elongation (Downs, 1962).

Rooting of cuttings, especially those of woody plants, is dramatically accelerated by long days (Piringer, 1961). Although not a great deal is known about how photoperiod affects the root initials it seems certain that it is a direct effect and not due to a promotion of shoot growth.

PHOTOPERIOD CONTROL

Garner and Allard (1920) found they could inhibit flowering of short day plants and promote flowering in long day ones during the short days of winter by applying supplementary artificial light after sunset. Extending the natural day with light of intensities as low as 50–100 lx gave essentially the same formative effects as were produced by the higher intensities of the additional daylight hours of the summer months. If given continuously throughout the dark period light levels as low as 2 lx may be enough to control flower initiation (Borthwick and Parker, 1938). Considering the low energy requirement it should surprise no one that stray light entering greenhouses or plant growth chambers can cause problems with flower production in short day plants. Lights from automobiles have inhibited flowering of plants grown in greenhouses located at curves of moderately traveled roads and research has been hampered by the location of street lights near greenhouses on the university campus. Since many plant growth chambers can hardly be considered light tight they should be located as far as possible from building lights that remain on continuously to provide emergency illumination.

Photoperiod control and the duration of the high intensity period do not necessarily induce the same physiological response. Moreover, where the high intensity light system is used to provide both long and short days it is difficult, if not impossible, to separate the photoperiod effect from the results of an increased period of photosynthesis. For example, flax will flower much earlier on natural summer days than it will with 8 h of sunshine plus 8 h of low intensity supplemental light. Thus, flax is not photoperiodically sensitive but simply responds to greater levels of photosynthate. In some phytotrons plants are grown in the greenhouses for 8 or 9 h then moved to growth rooms where the high intensity light is used to extend the day. When such methods are used the biological response cannot with certainty be attributed to long days as a photoperiodic phenomenon. Many so-called photoperiod responses may therefore need to be re-examined.

Garner and Allard (1931) showed quite clearly that the biological action takes place during the dark period. Thus, if the dark period is interrupted by an exposure to light the action is reset and must begin again during the period of darkness following that interruption. If the remaining dark period is too short to allow completion of the process the result is the same as if the plants had received long days and short nights. The dark period interruption using relatively low intensity light can therefore be used to provide a long day effect while minimizing the difference in total energy between the long and short day treatments.

In the S.E.P.E.L. long days are obtained by using a 3 h interruption near the middle of the long dark period that follows a basic 9 h high intensity light period. The basic light period of 9 h was selected only because many plants were moved each day to obtain a wider range of environmental conditions from the number of controlled-environment rooms available. Many plants like *Xanthium pensylvanicum* would require only a minute or two of light as a dark period interruption, especially at light levels of 20–30 hlx, whereas *Chrysanthemum*, *Hyoscyamus* and others would require much longer exposures. The 3 h interruption period was selected because: (1) it was successful in providing a long day effect over many years of use by the photoperiod project at Beltsville (Downs, 1956); and (2) the photomorphogenic effect is less than when the day is extended for several hours.

Some biologists do not wish to use the dark period interruption technique because they consider it "unnatural". More important are the legitimate applications such as soybean research, where daylengths intermediate between short and long ones are a necessary and valid part of the program. In these cases the base high intensity light period is lengthened with low intensity light. For no particular reason that we can determine, daylengths have almost always been *extended* with low intensity light—a procedure that demands power during the period of peak power usage. If the photoperiods were lengthened by adding low intensity light at the beginning of the light period, power would usually be used at a more economical time of day. Moreover the photomorphogenic effects of the extending light source would be largely avoided. However, supplementary light at the beginning of the day could result in an effectively shorter day than when photoperiods are extended in the usual way by the very lack of that photomorphogenic effect.

True photoperiodism is a response to changes in daylength through activation of the phytochrome pigment system: the same system that controls germination of light-sensitive seeds. Basic questions of photoperiodism include why some plants require a much longer light period for the interruption than others. The answer lies in the behavior of the P_{fr} form of phytochrome, especially the dark reversion rate: flowering of short day plants is prevented by converting the P_r form of phytochrome to P_{fr} by the light interruption. Present knowledge of dark reversion, the thermal conversion of P_{fr} to P_r, would probably benefit by further study.

The time that can elapse between the red and far-red irradiations without a loss of reversibility seems to be a measure of dark reversion rate. In *Paulownia* seeds P_{fr} remains active for more than 24 h (Borthwick *et al.*, 1964) and in bean P_{fr} seems to remain undiminished throughout the night (Downs *et al.*, 1957). Reversibility of cocklebur flowering, however, was found to be lost about 45 min after the red irradiation (Downs, 1956), and with *Chenopodium album* the reaction had proceeded beyond far-red control in less than 1 min, so that reversibility could not be demonstrated. However, since the phytochrome photoreaction is equally effective at low temperatures which reduce the dark reversion rate, far-red reversibility of *C. album* could be shown by simply reducing the temperature during the red irradiation. In plants such as *Chrysanthemum* P_{fr} may be maintained at an effective level by brief periods of light providing the dark period between successive irradiations is short enough for dark reversion not to deplete the P_{fr} level below the effective amount. Thus while 8 min of light is without effect, the same 8 min distributed over 4 h in cycles of 1 min every 30 min inhibits flowering (Borthwick and Cathey, 1962).

Another basic question arises from the action spectra which indicates that P_{fr} acts to prevent flowering of short day plants but promotes flowering of long day ones. The explanation proposed some years ago by Borthwick *et al.* (1956) is probably still as good as any (Fig. 16). The scheme is explained as follows. A flower-inducing substance must attain a certain minimum concentration and remain at that effective level for a sufficient period to produce an effect. If the concentration is too high a quenching effect occurs that limits effectiveness. Thus, a short day plant would, during the long dark period, slowly acquire the limiting level and the critical photoperiod would be a measure of the time required to attain that level. The differences in critical photoperiods may also include the time required for P_{fr} to be converted in the dark to P_r—a condition that may be necessary for the manufacture or the build-up of the flower-inducing substance. Obviously then, a dark period interruption would create P_{fr} which in turn stops production of the flower-inducing material and the critical level is not met or not maintained for a sufficient period to bring about the necessary changes.

Long day and indeterminant plants would enter the dark period with the effective substance within the critical limits. An increase of the material during a long night would soon exceed the critical limits and the plants would fail to flower. A dark period interruption would, by the production of P_{fr}, follow the same patterns as with short day plants, inactivating all or a substantial portion of the inducing material to return it within the operational concentration limits. The indeterminant types would have some method of limiting the flower-inducing substance to keep it within the limits: an inhibitor system, lack of substrate, or other metabolic control.

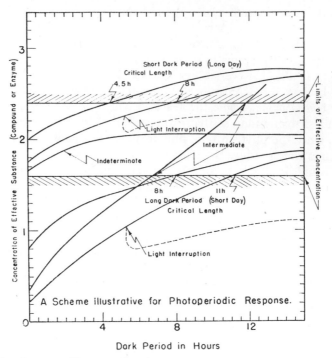

FIG. 16. A scheme to illustrate the photoperiod response (Borthwick *et al.*, 1956).

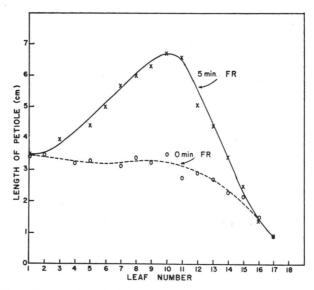

FIG. 17. Effect of far-red radiation at the close of 8 h light periods on petiole elongation in *Hyoscyamus niger*.

Granted such an explanation is risky, if for no other reason than that it is based on a hypothetical flower-inducing substance. However, one may recall experiments that clearly show that some kind of flower-controlling stimulus moves from the leaves to the site of flower differentiation. Also it seems fully established that P_{fr} needs to be at very low levels near the middle of the dark period for flowering of short day plants and at very high levels for promotion of flowering of long day ones. These P_{fr} levels are set by dark periods of proper duration to maintain an effective level of P_{fr} or by the light used to interrupt the dark periods.

Schneider *et al.* (1967) suggest that a high energy reaction (HER) is the major system controlling flowering in long day plants like *Hyoscyamus niger* and that the HER is neither a manifestation of photosynthesis or of phytochrome action. This may be true but the evidence is weak; it consists mainly of the fact that the response was not saturated by 17 hlx of incandescent light given as a 6 h dark period interruption. The flowering response in long day plants is an additive effect of flower initiation and flower development. Such additive effects can create confusing results, especially if the two responses are under a different type of control. Photocontrol of a typical long day plant occurs as follows. If phytochrome is placed predominantly in the P_r form at the close of a non-inductive photoperiod there is no measurable effect on flowering. Phytochrome is present and functioning, however, as illustrated by the change in petiole length of *Hyoscyamus* (Fig. 17); a fact already demonstrated by Wassink and Sytsema (1958). However, if the flower initiation processes are started by a pretreatment of several inductive cycles then the same brief exposure to far-red accelerates flowering and the typical phytochrome reversibility is demonstrated (Table XVII). Moreover, the reproductive system becomes

TABLE XVII

Reversibility of far-red promotion of flowering of *Hyoscyamus niger* by a subsequent red irradiation

Pretreatment[a] photoperiod (h)	Treatment[b] far-red (min)	Red (min)	Stem length (mm)	Stage of flowering
8	0	0	0	0·0
8	5	0	0	0·0
16	0	0	13	3·0
16	5	0	43	6·0
16	5	5	15	3·8

[a] 10 days of pretreatment
[b] 10 days of treatment; irradiations at the close of each 8 h day

sensitive to a far-red irradiation prior to any visible signs of differentiation. P_{fr} therefore is required for induction of the flowering processes, but once they are started P_{fr} is no longer required and may in fact be inhibitory. De Lint (1960) using experimental methods more like those of Schneider *et al.* (1967), reached essentially the same conclusion that "periodic influences operated through initiation with some elongation response but that formative influences entered the system via elongation". Thus, in practice a pure red light source is often less effective in photoperiodic flower control than one emitting about a 1:1 ratio of red and far-red. Examples are many and characteristically involve plants where the flowering process includes elongation of a scape or flower stalk; a feature of many long day plants.

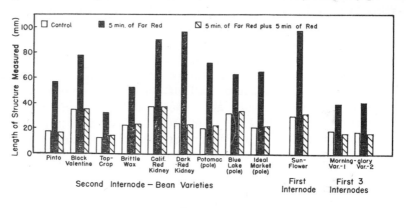

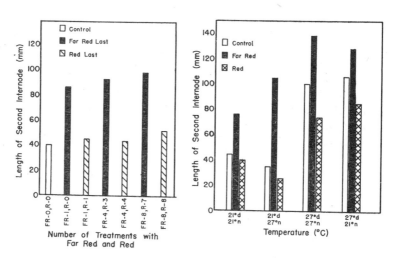

FIG. 18. Effect of red and far-red irradiations at the beginning of the dark period on internode elongation of bean, sunflower, and morning glory plants grown in controlled-environment rooms (Downs *et al.*, 1957).

TABLE XVIII

Effect of photoperiod and kind of supplemental light on the flowering behavior of Little Club wheat and Colsess I barley

Photo-period	Suppl. light	Period to heading		Stem length		Spike length		Spike weight		Fertile grains per spike	
		Days wheat	Days barley	cm wheat	cm barley	cm wheat	cm barley	g wheat	g barley	no. wheat	no. barley
8	none	—	—	—	—	—	—	—	—	—	—
12	Incandescent	101	107	90	62	70	57	0·74	0·34	30	7
12	Fluorescent	114	119	84	53	56	55	0·46	0·26	23	1
16	Incandescent	74	54	96	62	42	40	0·97	0·88	26	13
16	Fluorescent	92	98	80	57	55	45	0·37	0·22	26	5

It seems quite clear that with *Hyoscyamus*, wheat, millet, dill, sugar beet and others that P_{fr} is necessary to begin the initiation processes. Once the processes are begun an elongation response becomes involved. Experiments show that P_{fr} inhibits elongation in either dark-grown or light-grown plants (Downs, 1955; Downs et al., 1957) so it is not unreasonable to expect P_{fr} to inhibit scape elongation. Scape elongation and rate of flower development in some plants at least are related, and for most rapid flowering P_{fr} needs to be present for initiation but absent for development.

Short day plants or those that do not require scape or flower stalk elongation as a part of the flowering process react in a similar way, but it can only be detected under rather unusual conditions. In *Xanthium* plants on 1·5–2 h photoperiods for example, P_{fr} does not act during the first 6–8 h of darkness. It promotes flowering if present during the eighth to tenth hours. After the tenth hour it is no longer present, but if reintroduced by a red irradiation P_{fr} will then inhibit flowering (Borthwick and Downs, 1964); Mancinelli and Downs, 1967).

In practice, therefore, it is not surprising that the incandescent-filament lamp which converts 50–60% of the phytochrome to P_{fr} is frequently more effective than fluorescent ones which convert 70–80%. Wheat and barley spike emergence occurs earlier with incandescent supplementary light and the spikes are considerably larger (Table XVIII). With *H. niger* plants incandescent supplemental light not only accelerates anthesis but also has a marked influence on the percentage of flowers that develop into a fruit, (Table XIX) apparently due to some control of filament elongation.

TABLE XIX

Reproduction of *Hyoscyamus niger* as affected by source of supplementary light

Total light period (h)	Supplemental light Type	Duration (h)	Stem length (cm)	Days to anthesis	Fruit set (%)
8	none	0	0·2	veg.	—
16	Incandescent	8	42	27	66
16	Fluorescent	8	34	36	12

PHOTOMORPHOGENESIS

The regulatory effects of light on vegetative development, recently reviewed by Mohr (1972), are partly due to the phytochrome system and partly to a high energy reaction (HER). The phytochrome system controlling morphogenesis is typically characterized by saturation at relatively low intensities, the reciprocity rule holding over a relatively wide range of intensities and times, and of course, the reversible nature of the response by red and far-red radiant energies. The effects of high intensity radiation as a regulator of plant growth and development have been studied for many years (Wassink and Stolwijk, 1956). Mohr (1957) used the term HER to describe a photomorphogenic reaction requiring greater energies and having a different action spectrum from phytochrome. The main characteristics of this HER are the blue–far-red maxima, the requirement for long exposures to radiation and the fact that the system may act where phytochrome apparently does not. The two photoreactions, HER and phytochrome, are reported to act synergistically in some cases and antagonistically in others. To illustrate these photobiological effects one may observe that a bean plant grown in complete darkness has a very long hypocotyl, the leaves are small and folded and a plumular hook is present. Usually, the tissue is devoid of chlorophyll. Exposure to low light levels for only a few minutes per day straightens the plumular hook, inhibits hypocotyl elongation and causes the leaves to unfold and expand. These responses are most effectively induced by red radiant energy and are reversed by a subsequent exposure to far-red: in other words, a typical phytochrome reaction. Thus, the bean hypocotyl which reaches 30 cm in darkness will be reduced to about 20 cm by the activity of P_{fr} form of phytochrome. The potential inhibition, however, is 10 cm; a length obtained at 150 hlx of fluorescent–incandescent growth chamber light for 8 h per day. The second reaction might be due to the HER.

The HER is complex and difficult to analyze. Some evidence suggests the HER is a manifestation of phytochrome action and some indicates it is a separate photosystem. Borthwick et al. (1969) have extended Hartmann's (1966) model to account for the HER in terms of phytochrome and they account for many, but perhaps not all, of the observed plant responses.

The effects of the HER are sometimes controlled by a subsequent low energy, typical phytochrome, reaction. Moreover, the blue light effects may not be actuated in the same way as those produced by red or far-red (Meijer, 1968). However, in practice the development of plants grown in high intensity light is susceptible to phytochrome control. At the close of the light period the percent transformation of phytochrome to P_{fr} largely determines internode length (Fig. 18). The degree of the response is regulated by the number of hours of darkness in which the plants remain with a low P_{fr} level. Thus, at the end of the day if P_{fr} is removed by an irradiation with far-red light, internodes become elongated and the elongation is greater on a 20 h dark period than on a 10 h one.

Photomorphogenesis is not simply a laboratory response used by photobiologists in basic studies. In controlled environment facilities the HER and phytochrome become practical research tools, and if used improperly they can result in undesirable or abnormal plant growth. In practice,

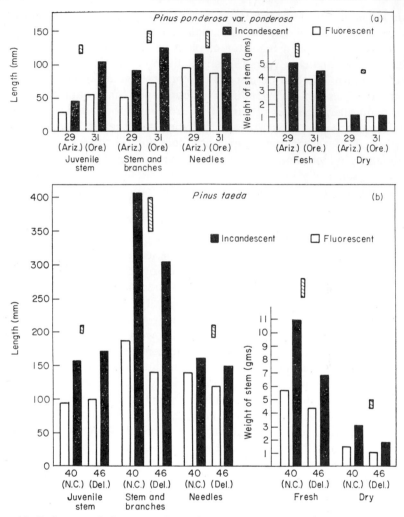

FIG. 19. Influence of the type of supplemental light used to extend the day on vegetative growth of pine.

ending the day with light from incandescent lamps reduces P_{fr} to a lower level than it was during the light period; which in most cases would be either the natural light of the greenhouse or the fluorescent–incandescent light of the controlled-environment room. Ending the day with fluorescent light, which produces a R–FR ratio from six to twenty times greater than that of the main light period, would greatly increase P_{fr}. It is quite possible then to produce plants with abnormally short internodes by ending the day with light from fluorescent lamps or excessive elongation may occur by ending the day with incandescent light (Table XX).

TABLE XX

Effect of light quality at the close of the light period on growth of tobacco
seedlings

Variety	Light source[a]	Stem length (cm)	Fresh weight Leaves (g)	Fresh weight Stem (g)	Fifth leaf Length (cm)	Fifth leaf Width (cm)
C-319	Fluorescent	6·3	10·3	1·1	9·7	16·2
	Incandescent	13·7	13·2	4·2	10·0	19·7
NC-2326	Fluorescent	5·7	11·9	2·0	9·0	15·8
	Incandescent	10·0	15·0	3·7	10·2	19·5
C-254	Fluorescent	8·4	12·4	2·6	8·9	17·0
	Incandescent	13·2	16·6	4·1	9·4	18·9
NC-98	Fluorescent	6·5	10·8	2·1	9·9	17·0
	Incandescent	12·8	12·3	4·3	9·1	18·0
C-298	Fluorescent	8·5	11·7	2·4	9·1	16·1
	Incandescent	14·9	15·3	4·3	9·8	18·8

[a] For 30 min at the end of 8½ h of Fluorescent + Incandescent at 430 hlx

TABLE XXI

Effect of photoperiod and kind of supplemental light on growth of *Acer rubrum*,
Catalpa bignonioides and *Hibiscus syriacus*

Photo-period	Supplemental light	*Acer rubrum* Nodes	*Acer rubrum* Stem length (cm)	*Catalpa bignonioides* Nodes	*Catalpa bignonioides* Stem length (cm)	*Hibiscus syriacus* Nodes	*Hibiscus syriacus* Stem length (cm)
8	none	9	32	2	2	3	3
16	Incandescent	22	117	6	21	14	38
16	Fluorescent	17	63	6	11	12	26

Woody plants whether deciduous (Table XXI) or evergreen (Fig. 19)
also respond to the P_{fr} ratio at the close of the light period. Some plant
parts such as stems and branches are more sensitive to the light source
than other parts such as needles or leaves. Plant habit may also be influenced by the light system and observations like those of Mapes (Williams
and Ungar, 1972) that *Suaeda depressa* grew erect in the greenhouse but
decumbent in plant growth chambers could be such a morphogenic effect
produced by the R:FR ratio.

USING ARTIFICIAL LIGHT FOR PLANT GROWTH

Excellent and "normal" plant development can be obtained using cool white fluorescent and incandescent-filament lamps in a 3:1 installed wattage ratio. With 1500 mA fluorescent lamps commercially called Power Groove, Very High Output (VHO) or Super High Output (SHO), 430–480 hlx can be maintained by a lamp changing schedule requiring a modest amount of labor time.

As a general rule the more hours of high intensity light there are, the greater is the dry weight production. The increase in plant growth, however, may not be proportional to the increase in energy. Moreover, long periods of light used as a standard procedure may create problems by prematurely inducing flowering in long day plants or retarding flowering of short day ones. The overall most adaptable high intensity light period is 8 or 9 h. This light duration would enable plants to be moved between chambers or to photoperiod rooms each day thereby increasing the overall flexibility of a group of controlled-environment facilities.

The incandescent–fluorescent system can be programmed to control morphogenesis by maintaining different levels of the P_{fr} form of phytochrome. Not only should the two light sources be programmed separately but for maximum flexibility some of the fluorescent lamps, about 20% of the installed wattage, should be controllable separately from or together with the remainder of the lamps.

Other kinds of light sources, especially the high intensity discharge lamps, are being used in plant growth chambers, However, very few comparative biological response data are available, so such light sources need to be used with considerable forethought. Some investigators have suggested that with a high level of PAR, spectral quality becomes much less important. This may be so if all the required wavelengths are present but it seems unlikely that a light source devoid of some important waveband such as 700–800 nm would be satisfactory at any intensity. McCree (1972) clearly showed that the Lucalox lamp has the greatest photosynthetic efficiency: 9% better than the cool white fluorescent. These data are in terms of PAR. The Lucalox may produce more PAR per watt but that alone does not mean plant growth will be acceptable. We can, for example, also get high levels of PAR with mercury lamps but again it is illogical to assume that acceptable plant growth will result. PAR alone is simply not enough. Like nutrition, the various components (spectral regions) must be balanced and that balance can result in success or failure to produce satisfactory plants. Unlike nutrition, however, the best balance of spectral regions has never been established. The balance provided by the usual fluorescent–incandescent system works biologically and while it very probably is not optimum the result in terms of plant growth is at least satisfactory. The same comments might also be made for a similar ratio of metal halide and incandescent lamps.

References

Annual Report 1971 (1972). Division of Plant Industry, C.S.I.R.O., Canberra, Australia.

Allphin, W. (1959). "Primer of Lamps and Lighting". Chilton, Philadelphia.

Army, T. J. and Greer, F. A. (1967). Photosynthesis and crop production systems. *In* "Harvesting the Sun" (A. San Pietro, F. A. Greer and T. J. Army, eds), pp. 321–332. Academic Press, New York and London.

Balegh, S. E. and Biddulph, O. (1970). The photosynthetic action spectrum of the bean plant. *Plant Physiology* **46**, 1–5.

Biamonte, R. L. (1972). The effects of light intensity of the initiation and development of flower primordia and growth of geraniums. M.S. Thesis, North Carolina State University.

Bickford, E. D. and Dunn, S. (1972). Lighting for Plant Growth. Kent State University Press.

Biological Handbook II. Growth (1962) (P. L. Altman and D. S. Dittmer, eds). Federation of American Societies for Experimental Biology.

Bjorkman, O. and Ludlow, M. M. (1972). Characterization of the light climate on the floor of a Queensland rainforest. *Carnegie Institute Year Book* **71**, 85–94.

Bjorkman, O., Boardman, N. K., Anderson, J. M., Thorne, S. W., Goodchild, D. J. and Pyliotis, N. A. (1972). Effect of light intensity during growth of *Atriplex patula* on the capacity of photosynthetic reactions, chloroplast components and structure. *Carnegie Institute Year Book* **71**, 115–135.

Black, M. and Wareing, P. F. (1955). Growth studies in woody species. VII. Photoperiodic control of germination in *Betula pubescens*. *Physiologia Plantain,* **8**, 300–316.

Blackman, G. E. and Black, J. N. (1959). XI. A further assessment of the influence of shading on the growth of different species in the vegetative phase. *Annals of Botany*, **23**, 51.

Bohning, R. H. and Burnside, C. A. (1956). The effect of light intensity on rate of apparent photosynthesis in leaves of sun and shade plants. *American Journal of Botany* **43**, 557–561.

Borthwick, H. A. and Cathey, H. M. (1962). Role of phytochrome in control of flowering of *Chrysanthemum*. *Botanical Gazette* **123**, 155–162.

Borthwick, H. A. and Downs, R. J. (1964). Roles of active phytochrome in control of flowering of *Xanthium pensylvanicum*. *Botanical Gazette* **125**, 227–231.

Borthwick, H. A. and Parker, M. W. (1938). Effectiveness of photoperiodic treatments of plants of different ages. *Botanical Gazette* **100**, 245–249.

Borthwick, H. A. and Parker, M. W. (1940). Floral initiation in Biloxi soybeans as influenced by age and position of leaf receiving photoperiodic treatment. *Botanical Gazette* **101**, 806–817.

Borthwick, H. A., Parker, M. W. and Heinze, P. H. (1941). Effect of photoperiod and temperature on development of barley. *Botanical Gazette* **103**, 326–341.

Borthwick, H. A., Hendricks, S. B. and Parker, M. W. (1956). Photoperiodism. *In* "Radiation Biology". (A. Hollaender, ed.), vol. 3, pp. 479–517. McGraw-Hill, New York.

Borthwick, H. A., Hendricks, S. B., Schneider, M. J., Taylorson, R. B. and Toole, V. K. (1969). The high energy light action controlling plant responses and development. *Proceeding of the National Academy of Sciences* **64**, 479–486.

Borthwick, H. A., Toole, E. H. and Toole, V. K. (1964). Phytochrome control of *Paulownia* seed germination. *Israel Journal of Botany* **13**, 122–133.

Bretschneider-Herrmann, B. (1962). Das Phytotron in Rauschholzhausen. *Zeitschrift für Acker-und Pflanzenbau* **115**, 213–222.

Bretschneider-Herrmann, B. (1969). The phytrotron in Rauischholzhausen; technical details and experiences. *Phytotronique* **1**, 24–26.

Burnside, C. A. and Bohning, R. H. (1957). The effect of prolonged shading on the light saturation curves of apparent photosynthesis. *Plant Physiology* **32**, 61–63.

Burton, W. G. (1972). The response of the potato plant and tuber to temperature. *In* "Crop Processes in Controlled Environments" (A. R. Rees, K. E. Cockshull, D. W. Hand and R. G. Hurd, eds), pp. 217–235, Academic Press, London and New York.

Chandler, B. (1972). Toward a simple growth room design. *Phytotronique* **II**, 281–287.

Clayton, R. K. (1965). "Molecular Physics in Photosynthesis." Blaisdell, Waltham, Mass.

Clayton, R. K. (1970). "Light and Living Matter: The Physical Part." McGraw-Hill, New York.

Cumming, B. G. (1959). Extreme sensitivity of germination and photoperiodic reaction in the genus *Chenopodium*. *Nature* **184**, 1044–1045.

De Lint, P. J. A. (1960). An attempt to analysis of the effect of light on stem elongation and flowering in *Hyoscyamus niger*. *Mededelingen van de Landbouhogeschool te Wageningen* **60**, 1–59.

Downs, R. J. (1955). Photoreversibility of leaf and hypocotyl elongation of dark grown Red Kidney bean seedlings. *Plant Physiology* **30**, 468–472.

Downs, R. J. (1956). Photoreversibility of flower initiation. *Plant Physiology* **31**, 279–284.

Downs, R. J. (1962). Photocontrol of growth and dormancy in woody plants. *In* "Tree Growth" (T. T. Kozlowski, ed.), Ronald Press, New York.

Downs, R. J. (1964). Photocontrol of germination of seeds of the Bromeliaceae. *Phyton* **21**, 1–6.

Downs, R. J. and Bailey, W. A. (1967). Control of illumination for plant growth. *In* "Methods in Developmental Biology" (F. H. Wilt and N. K. Wessels, eds), pp. 635–645. Thomas Crowell, New York.

Downs, R. J. and Borthwick, H. A. (1956). Effects of photoperiod on growth of trees. *Botanical Gazette* **117**, 310–326.

Downs, R. J., Borthwick, H. A. and Piringer, A. A. (1958). Comparison of incandescent and fluorescent lamps for lengthening photoperiods. *Proceedings of the American Society of Horticultural Science* **71**, 568–578.

Downs, R. J., Hendricks, S. B. and Borthwick, H. A. (1957). Photoreversible control of elongation of Pinto beans and other plants. *Botanical Gazette* **118**, 119–208.

Downs, R. J., Norris, K. H., Bailey, W. A. and Klueter, H. H. (1964). Measurement of irradiance for plant growth and development. *Proceedings of the American Society of Horticultural Science* **85**, 663–671.

Downs, R. J. and Piringer, A. A. (1955). Differences in photoperiodic responses of everbearing and June-bearing strawberries. *Proceedings of the American Society of Horticultural Science* **66**, 234–236.

Downs, R. J. and Piringer, A. A. (1958). Seed germination in the Bromeliaceae. *Bromeliad Society Bulletin* **8**, 36–38.

Dunn, S. and Went, F. W. (1959). Influence of fluorescent light quality on growth and photosynthesis of tomato. *Lloydia* **2**, 302–324.

Elenbaas, W. (1959). "Fluorescent Lamps and Lighting." Philips Technical Library.

Evans, L. T., Hendricks, S. B. and Borthwick, H. A. (1965). The role of light in suppressing hypocotyl elongation in lettuce and petunia. *Planta* **64**, 201–218.

Friend, D. J. C., Fischer, J. E. and Helson, V. A. (1963). The effect of light intensity and temperature on floral initiation and inflorescence development of Marquis wheat. *Canadian Journal of Botany* **41**, 1663–1674.

Gaastra, P. (1959). Photosynthesis of crop plants as influenced by light, carbon dioxide, temperature and stomatal diffusion resistance. *Mededelingen van de Landbouwhogeschool te Wageningen* **59**, 1–68.

Gaastra, P. (1964). Some comparisons between radiation in growth rooms and radiation under natural conditions *Phytotronique* **I**, 45–53.

Garner, W. W. and Allard, H. A. (1920). Effect of the relative length of day and night and other factors of the environment on growth and reproduction in plants. *Journal of Agricultural Research* **18**, 553–706.

Garner, W. S. and Allard, H. A. (1931). Effects of abnormally long and short alternations of light and darkness on growth and development of plants. *Journal of Agricultural Research* **42**, 829–851.

Hartman, K. M. (1966). A general hypothesis to interpret "high energy phenomena" of photomorphogenesis on the basis of phytochrome. *Photochemistry and Photobiology* **5**, 349–366.

Hatch, M. D., Osmond, C. B. and Slayter, R. O. (1971). "Photosynthesis and Photorespiration." Wiley Interscience, New York.

Hendricks, S. B., Toole, E. H., Toole, V. K and Borthwick, H. A. (1959). Photocontrol of plant development by the simultaneous excitations of two interconvertible pigments. III. Control of seed germination and axis elongation. *Botanical Gazette* **121**, 1–8.

Hoover, W. H., Johnston, E. S. and Brackett, F. S. (1933). Carbon dioxide assimilation in a higher plant. *Smithsonian Miscellaneous Collections* **87**, No. 16.

Hsiao, A. I. and Vidaver, W. (1971). Water content and phytochrome-induced potential germination responses in lettuce seeds. *Plant Physiology* **47**, 186–188.

IES (1968). Lighting Handbook, (J. E. Kaufman, ed.). Illuminating Engineers Society, New York.

Isikawa, S. (1954). Light sensitivity against the germination. I. Photoperiodism of seeds. *Botanical Magazine (Tokyo)* **67**, 51–56.

Jividen, G. M., Downs, R. J. and Smith, W. T. (1970). Plant growth under high intensity discharge lamps. Paper No. 70–824 Ann. Meeting Am. Soc. Agri. Eng.

Johnson, H. W., Borthwick, H. A. and Leffel, R. C. (1960). Effects of photoperiod and time of planting on rates of development of the soybean in various stages of the life cycle. *Botanical Gazette* **122**, 77–95.

Kahn, A. (1960). An analysis of dark-osmotic inhibition of germination of lettuce seed. *Plant Physiology* **35**, 1–7.

Kamen, M. D. (1963). "Primary Processes in Photosynthesis." Academic Press, London and New York.

Kawarada, A. and Shibata, K. (1972). Phytotron: Institute of Physical and Chemical Research. *Japanese Society for Environmental Control in Biology*, Supplement, 87–93.

Klueter, H. H., Bailey, W. A., Bolton, P. N. and Krizek, D. T. (1971). Xenon light and temperature effects on photosynthesis in cucumber. *ASAE Paper* No. 71–935.

Langhans, R. W. and Maginnes, E. A. (1962). Temperature and light. *In* "Snapdragons; a manual of the culture, insects and diseases and economics of snapdragons." (R. W. Langhams, ed.), New York State Flower Grower Association. Ithaca, N.Y.

Leafe, E. L. (1972). The relationship between crop physiology and analytical plant physiology. *In* "Crop Processes in Controlled Environments", pp. 157–175. (A. R. Rees *et al.*, eds). Academic Press, London and New York.

Ludlow, M. M. and Wilson, G. L. (1971). Photosynthesis of tropical pasture plants. II. Temperature and illuminance history. *Australian Journal of Biological Sciences* **24**, 1065–1075.

Mancinelli, A. L. and Downs, R. J. (1967). Inhibition of flowering of *Xanthium pensylvanicum* by prolonged irradiation with far red. *Plant Physiology* **42**, 95–98.

Mancinelli, A. L., Yanev, Z. and Smith, P. (1967). Phytochrome and seed germination. I. Temperature dependence of relative P_{fr} levels in the germination of dark germinating tomato seeds. *Plant Physiology* **42**, 333–357.

Matsui, T., Aiga, I., Eguchi, H. and Asakawa, F. (1971). Biological studies on light quality in environment control. II. Biological spectrograph: with special reference to operational characteristics of the instrument. *Environmental Control in Biology* **9**, 111–118.

McCree, K. J. (1972). Test of current definitions of photosynthetically active radiation against leaf photosynthesis data. *Agriculture and Meteorology* **10**, 433–453.

Meijer, G. (1959). The spectral dependence of flowering and elongation. *Acta Botanica Neerlandica* **8**, 189–246.

Meijer, G. (1968). Rapid growth inhibition of Gherkin hypocotyls in blue light. *Acta Botanica Neerlandica* **17**, 9–14.

Mitchell, K. J. (1972). The new phytotron in New Zealand. NSF–UNESCO–SEPEL Symposium on Phytotronics, Durham, North Carolina.

Mohr, H. (1957). Der Einfluss monochromatischer Strahlung auf das Langenwachstum des Hypocotyls und auf die Anthocyanbildung bei Keimlingen von *Sinapsis alba. Planta* **49**, 289–405.

Mohr, H. (1972). "Lectures on Photomorphogenesis." Springer-Verlag, Berlin.

Mohr, H. and Schoser, G. (1959). Eine Interferenzfilter Monochromatorlage fur Photobiologische Zwecke. *Planta* **53**, 1–17.

Monk, G. S. and Ehret, C. F. (1956). Design and performance of a biological spectrograph. *Radiation Research* **5**, 88–106.

Nakayama, S. (1958). Studies on the dark process in the photoperiodic response of *Pharbitis* seedlings. *Scientific Reports, Tohoku University Biological Series* **24**, 37–183.

Nitsch, J. P. (1972). Phytotrons: past achievements and future needs. *In* "Crop Processes in Controlled Environments" (A. R. Rees *et al.*, eds), pp. 33–56. Academic Press, London and New York.

Norris, K. H. (1968). A spectrograph for action spectra studies in the 400–800 nm region. *Transactions ASAE* **11**, 407–408.

Nutile, G. E. (1945). Inducing dormancy in lettuce seed with coumarin. *Plant Physiology* **20**, 433–442.

Parker, M. W. and Borthwick, H. A. (1949). Growth and composition of Biloxi soybean grown in a controlled environment with radiation from different carbon-arc sources. *Plant Physiology* **24**, 345–358.

Parker, M. W., Hendricks, S. B., Borthwick, H. A. and Scully, N. J. (1946). Action spectrum for the photoperiodic control of floral initiation of short day plants. *Botanical Gazette* **108**, 1–26.

Piringer, A. A. (1961). Photoperiod, supplemental light and rooting of cuttings. *Proceedings 2nd Annual Meeting Western Plant Progagators Conference.*

Piringer, A. A. and Borthwick, H. A. (1955). Photoperiodic responses of coffee. *Turrialba* **5**, 72–77.

Piringer, A. A. and Cathey, H. M. (1960). Effect of photoperiod kind of supplemental light and temperature on the growth, and flowering of petunia plants. *Proceedings of the American Society of Horticultural Science* **76**, 649–660.

Piringer, A. A. and Downs, R. J. (1960). Effects of photoperiod and kind of supplemental light on the growth of *Theobroma cacao. Proceedings 18th Inter-American Cacao Conference Trinidad*, pp. 82–90.

Piringer, A. A., Downs, R. J. and Borthwick, H. A. (1958). Effects of photoperiod on Rauwolfia. *American Journal of Botany* **45**, 323–326.

Piringer, A. A., Downs, R. J. and Borthwick, H. A. (1963). Photocontrol of growth and flowering of *Caryopteris. American Journal of Botany* **50**, 86–90.

Rabinowitch, E. and Govindjee (1969). "Photosynthesis." John Wiley, New York.

San Pietro, A., Greer, F. A. and Army, T. J. (eds). (1967). "Harvesting the Sun: Photosynthesis in Plant Life." Academic Press, New York and London.

Schneider, M. J., Borthwick, H. A. and Hendricks, S. B. (1967). Effects of radiation on flowering of *Hyoscyamus niger. American Journal of Botany* **54**, 1241–1249.

Seliger, H. H. and McElroy, W. D. (1965). "Light: Physical and Biological Action." Academic Press, New York and London.

Siegelman, H. W., Turner, B. C. and Hendricks, S. B. (1966). The chromophore of phytochrome. *Plant Physiology* **41**, 1289–1292.

Sweet, G. B. and Wareing, P. F. (1966). Role of plant growth in regulating photosynthesis. *Nature* **210**, 77–79.

Thomas, M. D. and Hill, G. R. (1937). The continuous measurement of photosynthesis, respiration and transpiration of alfalfa and wheat growing under field conditions. *Plant Physiology* **12**, 285–307.

Toole, E. H., Borthwick, H. A., Hendricks, S. B. and Toole, V. K. (1953). Physiological studies of the effects of light and temperature on seed germination. *Comptes Rendus de l'Association Internationale d'Essais de Semences*. **18**, 267–276.

van Wijk, W. R. (1963). "Physics of Plant Environment." North-Holland, Amsterdam.

Vince, D. and Stoughton, R. H. (1957). Artificial light in plant experimental work. *In* "Control of the Plant Environment" (J. P. Hudson, ed.), pp. 5–82, Butterworths, London.

Warren Wilson, J. (1972). Control of crop processes. *In* "Crop Processes in Controlled Environments" (A. R. Rees *et al.*, eds), pp. 7–30. Academic Press, London and New York.

Wassink, E. C. and Stolwijk, J. A. J. (1956). Effects of light quality on plant growth. *Annual Review of Plant Physiology* **7**, 373–400.

Wassink, E. C. and Sytsema, W. (1958). Petiole length reaction in *Hyoscyamus niger* upon daylength extension with light of narrow spectral regions as correlated with the length of the basic light period and upon night interruption with red and far-red radiation. *Mededelingen van de Landbouwhogeschool te Wageningen* **58**, 1–6.

Went, F. W. (1957). "Environmental Control of Plant Growth." *Chronica Botanica* **17**, Ronald Press, New York.

Williams, M. D. and Ungar, I. A. (1972). The effect of environmental parameters on the germination, growth and development of *Suaeda depressa*. *American Journal of Botany* **59**, 912–918.

Withrow, R. B. (1957). An interference filter monochromator system for the irradiation of biological material. *Plant Physiology* **32**, 355–360.

Withrow, R. B. and Price, L. (1953). Filters for isolation of narrow regions in the visible and near-visible spectrum. *Plant Physiology* **28**, 105–114.

Zelitch, I. (1971). "Photosynthesis, Photorespiration and Plant Productivity." Academic Press, New York and London.

CHAPTER IV

Gaseous Content of the Atmosphere

Carbon Dioxide

NORMAL CONCENTRATIONS

The essential nature of carbon dioxide for plant growth is an accepted fact. It is generally conceded that plant species may have different abilities to absorb CO_2 and thus have different compensation points of CO_2 utilization. Therefore, species like Norway maple fail to absorb CO_2 at concentrations below 150 ppm, whereas others like corn absorb CO_2 at levels below 10 ppm. Growth rates would be expected to decline as the CO_2 decreased and to stop completely near the compensation point. Most investigators would probably agree that plant growth is often limited by the CO_2 concentration under natural conditions. Considering these things it seems strange that in most so-called controlled-environment facilities carbon dioxide concentration is rarely measured and seldom is the CO_2 monitored continuously with an effort to maintain it at a constant level.

Most plant growth chambers have a make-up air system which owners assume is to maintain ambient CO_2 levels. Although the amount of new air introduced by such systems is rarely stated, it usually does not exceed 10% of the room volume per minute. One might assume that this air exchange rate is satisfactory because it must have been determined by experimentation and measurement. Unfortunately, such determinations are almost never made. Most of the make-up air systems are useless and have led many investigators into the erroneous conclusion that no CO_2 stress occurs in the chambers.

How much make-up air is necessary? Measurements in the phytotron show that a room of tobacco plants at the time they are topped will reduce the CO_2 level from 400 to 200 ppm within 30–40 min after the onset of the light period. Using the make-up air system, which in this case changes 2% of the room volume per minute, has no appreciable effect. When the make-up air was increased to 10–12% of the room volume per minute, the five-fold to sixfold increase only altered the CO_2 level by about 25%. Obviously a much higher rate is necessary and Morse (1963) calculated that at least 1·53 m^3 min^{-1} of new air would be needed for each m^2 of growing space. Thus, a common size growth chamber, 1·22 × 2·44 m, would require 4·56 m^3 min^{-1} of outside air—about 75% of the room volume per minute. In view of the large amounts required, it just is not practical to try to maintain ambient CO_2 levels in a growth chamber by a make-up air system.

If ambient CO_2 levels are to be maintained by gas injection from tanks or generators, the CO_2 concentration must be measured periodically. In view of the cost and maintenance of i.r. gas analyzers and associated equipment, one must ask what is the practical gain to be derived. In other words, does the apparently depleted CO_2 level have a real effect on plant growth and behavior? Certainly net photosynthesis decreases alarmingly at reduced CO_2 concentrations. Net photosynthesis of *Encelia californica*, for example, is only 32% as great at 100 ppm as at 300 ppm CO_2 (Mooney et al., 1966). Takamura (1966) showed a linear decrease in net photosynthesis of tobacco and tomato as the CO_2 levels dropped below 300 ppm, stopping completely at 80–100 ppm. However, in the controlled-environment room occupied by a single type of plant, growth and appearance may seem normal. Reported reductions in photosynthetic rates (Hesketh, 1963; Takamura, 1966) are not expressed by a corresponding decrease in dry weight (Raper and Downs, 1973). This is because the reduction of CO_2 content is brought about by the plants themselves absorbing CO_2 faster than the leakage plus ventilation rate can replace it. Since the CO_2 stress is self-imposed in the controlled-environment room it does not affect plant growth to the extent that would result if the low level were obtained by artificial means.

The effect of CO_2 stress on tobacco is more subtle than a simple reduction in dry weight. In the phytotron tobacco requires a sequential nutrition program and a simulated seasonal temperature progression if the plants are to resemble field-grown ones (Raper, 1971) and produce leaves that ripen properly. Progressing from bottom to top of a tobacco plant a variety of physical and chemical leaf characteristics appear. These differences form an orderly progression that describes the environmental events that occurred during the growth of the plant. The progression of leaf areas in CO_2-depleted plants is not representative of field-grown tobacco although the dry weight per unit area may be similar. Internodes are shorter than normal

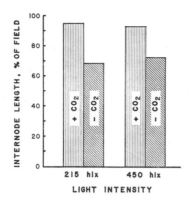

FIG. 1. Effect of CO_2 depletion on relative internode lengths of tobacco plants grown at two light intensities. (Raper and Downs 1973)

(Fig. 1) and an unnatural epinasty of leaves during maturation repeatedly occurs. CO_2 depletion also severely depresses reducing sugars in the upper leaves (Table I). Thus, while plant-suppressed CO_2 in the growth chamber does not give the same results as the artificially suppressed CO_2 in photosynthesis studies a stress does occur and it can have a detrimental effect on plant growth.

TABLE I

Effects of depleted CO_2 levels on accumulation of various components in tobacco leaves grown at three light intensities (Raper *et al.*, 1973)

| Type of compound[a] | Cumulative weights in leaves (g) | | | | | |
| | Maintained CO_2 | | | Depleted CO_2 | | |
	215 hlx	322 hlx	450 hlx	215 hlx	322 hlx	450 hlx
Leaf fresh weight	1212	1258	1282	1242	1212	1188
Carbohydrates						
Soluble	10·91	15·66	18·36	9·45	13·90	16·00
Starch	7·06	11·32	12·83	5·69	7·43	10·08
Reducing sugar	2·38	2·77	3·11	2·20	2·41	2·90
Non-reducing sugar	1·47	1·57	2·42	1·56	4·06	3·02

[a] Green leaves

In phytotron operations fairly large rooms are often occupied by more than one kind of plant. If half of a 9 m² room is occupied by corn plants the CO_2 level may drop below 100 ppm. Other kinds of plants such as geraniums and marigolds placed in the room may, because of their higher compensation points, turn chlorotic and fail to grow.

Although the data are not extensive, they certainly indicate that maintenance of ambient CO_2 concentrations is desirable and in many cases imperative. As a result the phytotron was modified to include a multi-channel, continuous flush, CO_2 monitor. Air from all chambers on the line is pumped to the analyzer continuously, so distance does not affect the time required to balance and read out each point.

After investigating a few of the CO_2 control systems reported in the literature (Bailey *et al.*, 1970; Pettibone *et al.*, 1970; Tatemichi, 1970; Slack and Calvert, 1972) we decided on the simplest possible method. Gas from pressure-regulated commercial bottles is continuously fed into manually controlled flow meters during the light period. Since the environmental conditions are constant the major factors influencing CO_2 consumption are leaf area and age, both of which change slowly. Therefore, a room of large, actively growing plants like tobacco will show a variation of only about 25 ppm with an adjustment twice per week. A change in plant load or environmental conditions will, of course, require a change in CO_2 input to maintain the desired conditions. The system will not scrub out the temporary CO_2 rise caused by people entering the space to maintain the plant material but it is a practical and very reliable system. Gas consumption is not very high if reasonable care is taken to avoid leaks. With an average plant loading, one 60 lb cylinder of CO_2 supplies seven 1·22 × 2·44 m chambers for 5 weeks. Coleman grade CO_2 is not necessary but only the highest C. P. or commercial grade should be used. Any compressor oil in the CO_2 will act as an air pollutant so even though the best commercial grade gas is apparently nearly free of such oils we also use a line filter.

ENCHANCED CO_2 CONCENTRATION

Although few papers have been published to show the importance of maintaining ambient CO_2 concentrations in controlled-environment rooms, a large literature exists on the practical advantages of increased CO_2 levels. Variously called enhanced, supplemental or enriched CO_2 studies reported from about 1900 to 1970 were discussed in the ASAE symposium on Controlled Atmospheres for Plant Growth (1970), and a historical survey was presented by Daunicht (1963) and by Wittwer and Robb (1964). After reading a number of reports on effects of increased CO_2 concentrations the following comment is particularly impressive:

> "The experimental evidence in regard to the increase in the photosynthetic rate due to increasing the normal carbon dioxide supply of the atmosphere is rather confusing. This is due to the fact that many workers did not consider other factors that are known to play a part and to the fact that the interaction of numerous factors in this regard is not understood by the investigators today."
>
> (Miller, 1938).

Although Miller made these remarks many years ago they seem equally appropriate today; especially the part about the lack of understanding of the "interactions of numerous factors".

Much of the enhanced CO_2 work has considered its effects on net photosynthesis. As early as 1902 Brown and Escombe showed that if the partial pressure of CO_2 was changed from 1 to 6·6 the ratio of CO_2 absorbed by the leaves was 1:7·2. More recent data (Hesketh, 1963) indicate that a 233% increase in CO_2 results in a 95% increase in the net rate of photosynthesis for corn and 188% for tobacco. At 300 ppm corn is a more efficient user of CO_2 than tobacco but as the concentration increases the efficiency of tobacco increases more than that of corn. Net photosynthesis certainly varies between species and between varieties within a species (Zelitch, 1971; Ford and Thorne, 1967; Hesketh, 1963) but at least some of the apparent variation may be due to different influences of temperature (Murata and Iyama, 163; Warren Wilson, 1966), age (Hurd, 1968; Ludlow and Wilson, 1971) and experimental techniques.

The increase in net photosynthesis due to additional CO_2 is in large part dependent upon any change in optima of other environmental factors brought about by that increase. For example, it is widely accepted that CO_2 utilization at normal 300–400 ppm is to some extent influenced by light intensity (Gaastra, 1959; Hesketh, 1963; Hesketh and Moss, 1963; Ford and Thorne, 1967) and that the saturation light level is usually relatively low. The low saturation light intensities, however, are due in large part to experimental methods and may not be valid in practice. De Wit (1965), for example, shows net photosynthesis of a single leaf to be saturated at about 0·15 cal cm^{-2} min^{-1} of PAR, whereas the average leaf was not saturated at three times that PAR energy. At enhanced CO_2 levels the saturation intensity may change many fold. Soybean, for example, exhibits saturation at 215 hlx at normal CO_2 levels, but when the CO_2 is increased to 1600 ppm light saturation is not reached at 754 hlx (Brun and Cooper, 1967).

The idea that increasing the CO_2 concentration will substitute for low light levels is somewhat exaggerated. Klougart (1967) in fact, stated bluntly that substitution of low light by CO_2 cannot be done although he agreed that some compensation may be obtained.

Dry weight production should equal net photosynthesis. However, a direct relationship between net photosynthesis and yield of a particular part of the plant such as top dry weight, leaf area or fruit may not hold. Additional discrepancies may arise from the fact that net photosynthesis is often measured for a brief period under rather ideal conditions, whereas dry weight production is the accumulation of many days' growth under conditions usually less than optimum. Some self-shading would be unavoidable and since plants at high CO_2 concentrations are probably below saturation irradiances, the amount of shading could be increasingly important. Moreover, the net photosynthesis per unit of leaf decreases as the plants grow (Ford and Thorne, 1967). This may be reflected in the report by Hurd (1968) that the growth rate of young tomato plants in enhanced CO_2 atmospheres is initially high but falls to near the rate of the normal CO_2 plants.

Many apparent discrepancies between net photosynthesis and dry weight production arise from experimental techniques or the way we interpret them. Ford and Thorne (1967), for example, pointed out that in their work the difference between photosynthesis and assimilation may have been due to a 4–6°C warmer temperature during the photosynthesis measurements. Duncan and Hesketh (1968) presented data for high-altitude corn, showing that dry weight decreased markedly as the temperature rose from 15 to 20°C although net photosynthesis remained essentially constant. In this case the CO_2 utilization data were obtained from the youngest fully emerged leaf which is probably the most efficient one. The data did serve to illustrate the relative photosynthetic efficiencies of the different corn races. However, it would be highly unlikely that a good correlation with dry weight would result because leaf area also changed with temperature, the leaves were of different ages and some degree of shading was involved.

Some evidence exists that the effectiveness of enhanced CO_2 atmospheres is greater with seedlings than with larger plants. Dullforce (1967) stated that additional CO_2 was effective only for about the first 6 weeks in lettuce. Little effect was obtained on later growth or on hearting. Growth of tobacco seedlings, for example, is markedly accelerated by high CO_2 levels (Table II) and the increase is maintained for about 5 weeks. After that time growth

TABLE II

Effect of increased CO_2 concentration on growth of Coker 319 tobacco seedlings for 18 days

Response	3–400 ppm		10–1200 ppm		GH	
	Expt 1	Expt 2	Expt 1	Expt 2	Expt 1	Expt 2
Leaf length (cm)	2·1	3·1	5·2	5·2	1·3	1·2
Leaf width (cm)	1·8	2·5	4·0	3·9	1·1	0·9
Fresh weight (mg)	180	399	1290	1219	40	50
Dry weight (mg)	9	—	87	—	3	—

must slow because the plants at normal CO_2 concentrations catch up and at 8 weeks are nearly equal to those grown with enhanced CO_2. As explained by Thomas et al. (1973), huge tobacco plants do not occur as a result of enhanced CO_2; instead the growth seems to be accelerated at high CO_2 levels with all processes occurring at a more rapid rate inducing early senescence.

Several investigators (Madsen, 1968; J. D. Hesketh, personal communication, 1972) have remarked on the high level of starch in leaves of plants grown at enhanced CO_2 concentrations. A recent test with tobacco showed a tenfold increase in starch per unit of leaf grown at 1200 ppm CO_2. Reports that mention high starch also note the chlorotic condition of the leaves and the brittle, leathery or hard texture. In Madsen's (1968) work starch was eight times higher and the chlorophyll content was reduced from 1·5 to 0·7 mg g^{-1} fresh weight of plants grown at high CO_2. Apparently the faster-growing plants in the enhanced CO_2 atmosphere deplete the available nitrogen and cause the typical chlorosis and carbohydrate increase. If the nutrient levels are stepped up at the higher CO_2 concentration large starch accumulations do not occur nor are the plants chlorotic. However, large additional increases in growth do not occur. J. D. Hesketh (personal communication, 1972) reports that with cotton the high starch accumulation and chlorosis in enhanced CO_2 atmospheres is avoided simply by increasing the temperature from 31 to 35°C.

Nutrition requirements change with enhanced CO_2 concentrations and although such statements appear in the literature they are rarely accompanied by any recommendations. Wittwer (1967) shows a marked improvement in the response of lettuce to enhanced CO_2 by doubling the nitrogen rate. Specific experiments in the phytotron show that the nutrient solution concentration can be doubled when the CO_2 concentration is raised from 400 to 1200 ppm (Table III). However, Mattson and Widmer (1971) reported that fertilization of roses had to be decreased in enhanced CO_2 atmospheres to avoid undesirable increases in nitrate nitrogen. Phytotron work with radish root growth also shows the undesirable effect of increasing nutrition levels (Table III).

What happens when plants are grown at high levels of CO_2? Of course, net photosynthesis increases and photorespiration could decrease. In addition, transpiration may decrease due to stomatal closure and leaf temperatures could rise several degrees (Pallas and Bertrand, 1966). There can be no doubt that the final result can be greater yields and dry weight production. The problem is why final growth results are not consistent and predictable at least to the same degree as net photosynthesis. Kretchman and Howlett (1970) show a year-to-year variation of from 5·4 to 24·3% yield increases. Higher temperatures generally, but not always, reduced the benefit of higher CO_2 concentrations. Spring crops usually benefited more than fall crops. Since this work was done in greenhouses many hypotheses can be developed to explain the differences: variable irradiance, the time CO_2 levels could be maintained before high temperatures required ventilation, etc. Another factor, however, was introduced by Hughes and Cockshull (1971). Using controlled-environment chambers they noted the *Chrysanthemum* plants obtained in January showed a large final total and flower dry weight with high CO_2 and there was a strong interaction between CO_2 and irradiance. Plants obtained in September, however, failed to respond well to enhanced CO_2. Thus, the environmental conditioning of the plants prior to placing them in the growth chambers exerted a marked control over the response to increased CO_2 concentrations.

TABLE III

Effect of nutrient and CO_2 concentration on growth of soybean (22 days), tomato (32 days) and radish (18 days) at 26 °C and 430 hlx for 15 h per day in a fluorescent–incandescent growth chamber

Nutrient concentration[a]	CO_2 concentration	Dare soybean	Manapal tomato	Cherry Belle radish	Root weight (g) Cherry Belle radish
		Fresh weight tops (g)			
1/4	3–400	22·2	71·9	6·4	5·1
	10–1200	32·5	78·2	7·2	15·3
1/2	3–400	19·3	102·1	5·4	3·9
	10–1200	43·2	172·1	7·6	12·3
1	3–400	18·6	69·4	4·4	2·1
	10–1200	34·8	150·2	7·2	9·5
Age in days		22	32	18	18

[a] Half-strength is the NCSU nutrient equivalent of a half-strength Hoaglands solution. Twice daily application

Obviously much more research will be needed if we are to use enhanced CO_2 atmospheres effectively. It seems quite reasonable to assume that the majority of these studies will require controlled environment facilities. The controlled-environment facilities of the phytotron are quite adaptable to this type of research without major modification or special construction.

Oxygen

Our measurements in the phytotron indicate that oxygen levels in controlled-environment rooms do not change appreciably from normal levels. Since oxygen concentration must deviate considerably from that normally present before it affects respiration or photosynthesis (Zelitch, 1971), we have had no sound reasons for considering it as an environmental factor in practice.

Pollutants

Phytotoxic constituents in the atmosphere are not new but they have finally become so obvious as to attract attention. Phytotoxicants can be a problem in any part of the country and as Went (1957) pointed out, "leaf-damaging types of air pollution only increases and never decreases in the course of the years". Therefore, all areas used for plant research should have a system of filters to remove pollutants from the ventilation air.

Studies involving phytotoxic atmospheric components can be performed to excellent advantage in controlled-environment rooms. Special treatment roomettes like those designed by the Air Pollution Research Laboratory headed by W. W. Heck (Heck et al., 1968) should be used as treatment chambers. These units are equipped with charcoal filters for inlet air and a filter on the exhaust side to remove the pollutant. They act as typical roomettes in that the environment of the growth room serves as base line conditions (Fig. 2).

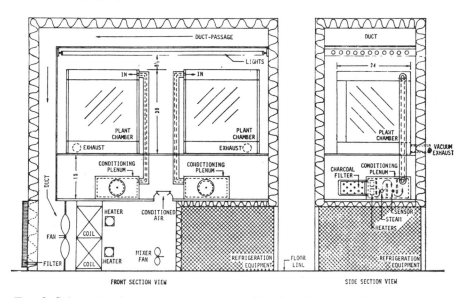

FIG. 2. Schematic of exposure roomettes used in air pollution studies (Heck et al., 1968).

The use of controlled-environment facilities is of importance to air pollution research because of the marked effect of many environmental factors on plant sensitivity. Several reviews (Middleton, 1961; Heck, 1968; Heggestad and Heck, 1971) have been prepared on the nature of plant responses to air pollutants so only a synopsis is necessary here.

Ozone is reported to be the most common air pollutant in North America and is probably best known as the cause of stipple of grapes and weather-fleck of tobacco. During the growth period sensitivity to ozone damage is greater if the exposure follows a long dark period or if the plants are grown with reduced nutrient levels. Low temperatures prior to exposure and water stress reduce sensitivity. Plants grown under low irradiance are more sensitive to ozone damage but less so to peroxyacetyl nitrate—the main causal agent of symptoms associated with smog. Environmental conditions during exposure also alter sensitivity. High irradiance and/or high humidity increase sensitivity of plants to ozone treatment while enhanced CO_2 reduces sensitivity. Of particular importance to phytotron users is the observation (Heck, 1972) that the injury due to air pollutants is less in controlled-environment room plants than in those grown in greenhouses or in the field. The decrease in sensitivity may be due to the greater growth rate and generally increased vigor of plants in CERs. Heck (1972) also noted that even under controlled conditions plants like Pinto bean show diurnal and seasonal differences in responsiveness to air pollutants. Generally the plants are less sensitive in winter and during the first hour of the light period. Sensitivity then increases to reach a maximum after about 3 h.

References

ASAE Symposium (1970). Controlled atmospheres for plant growth. *Transactions ASAE* **13**, 327–268.

Bailey, W. A., Klueter, H. H., Krizek, D. T. and Stuart, N. W. (1970). CO_2 systems for growing plants. *Transactions ASAE* **13**, 263–267.

Bailey, W. A., Klueter, H. H., Krizek, D. and Lui, R. C. (1972). The phyto-engineering laboratory. *Phytotronique* **II**, 91–107.

Brun, W. A. and Cooper, R. L. (1967). Effects of light intensity and carbon dioxide concentration on photosynthetic rate of soybean. *Crop Science* **7**, 451–454.

Coble, C. G. and Bowen, H. D. (1967). Oxygen concentration measurements in biological material. *Transactions ASAE* **10**, 325–326.

Daunicht, H. J. (1963). CO_2-Dungung, Entwicklung Heutiger stand eigene Versuchsergebnisse. *Acta Horticulturae* **2**, 86–96.

de Wit, C. T. (1965). Photosynthesis of leaf canopies. *Agricultural Research Report Wageningen* 663.

Dullforce, W. M. (1967). Analysis of the growth of lettuce in controlled environments with additional carbon dioxide. *Proceedings XVII International Horticultural Congress* **1**, 345.

Duncan, W. G. and Hesketh, J. D. (1968). Net photosynthetic rates, relative leaf growth rates and leaf numbers of twenty-two races of maize grown at eight temperatures. *Crop Science* **8**, 760–674.

Ford, M. A. and Thorne, G. N. (1967). Effect of CO_2 concentration on growth of sugar beet, barley, kale, and maize. *Annals of Botany* **31**, 629–644.

Gaastra, P. (1959). Photosynthesis of crop plants as influenced by light, carbon dioxide, temperature, and stomatal diffusion resistance. *Mededelingen van de Landbouwhogeschool Wageningen* **59**, 1–8.

Gilbert, S. G. and Shive, J. W. (1942). The significance of oxygen in nutrient substrates for plants. I. The oxygen requirements. *Soil Science* **53**, 143–152.

Heck, W. W. (1968). Factors influencing expression of oxidant damage to plants. *Annual Review of Phytopathology* **6**, 165–187.

Heck, W. W. (1972). Air pollution research on plants in phytotroms. *NSF–UNESCO–SEPEL Symposium*, Durham, North Carolina.

Heck, W. W., Dunning, J. A. and Johnson, H. (1968). Design of a simple plant exposure chamber. *National Center for Air Pollution Control* Publication APTD 68–6.

Heggested, H. E. and Heck, W. W. (1971). Nature, extent and variation of plant response to air pollutions. *Advances in Agronomy* **23**, 111–145.

Hesketh, J. D. (1963). Limitations to photosynthesis responsible for differences among species. *Crop Science* **3**, 493–496.

Hesketh, J. D. and Moss, D. N. (1963). Variation in the response of photosynthesis to light. *Crop Science* **3**, 107–110.

Hughes, A. P. and Cockshull, K. E. (1971). The variation in response to light intensity and carbon dioxide concentration shown by two cultivars of *Chrysanthemum morifolium* grown in controlled environments at two times of year. *Annals of Botany* **35**, 933–945.

Hurd, R. G. (1968). Effects of CO_2 enrichment on the growth of young tomato plants in low light. *Annals of Botany* **32**, 531–542.

Jividen, G. M. (1972). "Temperature and anaerobic effects upon normal germination of cotton and the differentiation of periods of chilling as a function of internal oxygen utilization." Ph.D. Thesis, North Carolina State University.

Klougart, A. (1967). A look ahead based on research on CO_2 and growth of horticultural plants in Europe. Proceedings *XVII International Horticultural Congress* **3**, 323–332.

Kretchman, D. W. and Howlett, F. S. (1970). CO_2 enrichment for vegetable production. *Transactions ASAE* **13**, 252–256.

Ludlow, M. M. snd Wilson, G. L. (1971). Photosynthesis of tropical plants. III. Leaf age. *Australian Journal of Biological Sciences* **24**, 1077–1087.

Madsen, E. (1968). Effect of CO_2 concentration on the accumulation of starch and sugar in tomato leaves. *Physiologia Plantarum* **21**, 168–175.

Mattson, R. H. and Widmer, R. E. (1971). Year round effects of CO_2 supplemented atmospheres on greenhouse rose production. *Proceedings ASHS* **96**, 487–488.

Middleton, J. T. (1961). Photochemical air pollution damage to plants. *Annual Review of Plant Physiology* **12**, 431–448.

Miller, E. C. (1938). "Plant Physiology." McGraw-Hill, New York.

Mooney, H. A., Strain, B. R. and West, M. (1966). Photosynthetic efficiency at reduced CO_2 tensions. *Ecology* **47**, 490–491.

Morse, R. N. (1963). Phytotron design criteria: engineering principles. *In* "Engineering Aspects of Environment Control for Plant Growth", pp. 20–39 C.S.I.R.O., Australia.

Murata, Y. and Iyama, J. (1963). Influence of air temperature upon photosynthesis of some forage and grain crops. *Proceedings of the Crop Science Society of Japan* **31**, 315–322.

Pallas, J. E. and Bertrand, A. R. (1966). Research in plant transpiration: 1963. *USDA Product Research Report* No. 89.

Pettibone, C. A., Matson, W. E., Pfeiffer, C. L. and Ackley, W. B. (1970). The control and effects of supplemental carbon dioxide in air supported plastic greenhouses. *Transactions ASAE* **13**, 259–262.

Raper, C. D. (1971). Factors affecting the development of flue-cured tobacco in artificial environments. III. Morphological behavior of leaves in simulated temperature, light-duration and nutritional progressions during growth. *Agronomy Journal* **63**, 848–852.

Raper, C. D. and Downs, R. J. (1973). Factors affecting the development of flue-cured tobacco in artificial environments. IV. Effects of carbon dioxide depletion and light intensity. *Agronomy Journal* **65**, 247–252.

Raper, C. D., Weeks, W. W., Downs, R. J. and Johnson, W. H. (1973). Chemical properties of tobacco leaves as affected by carbon dioxide stress and light intensity. *Agronomy Journal* **65**, 988–992.

Slack, G. and Calvert, A. (1972). Control of CO_2 concentration in glasshouses by use of conductimetric controllers. *Journal of Agricultural Engineering Research* **17**, 107–115.

Takamura, T. (1966). The effect of room ventilation on net photosynthesis rate. *Botanical Magazine (Tokyo)* **79**, 143–151.

Tatemichi, Y. (1970). On the automatic CO_2 controlling and recording system. *Environmental Control in Biology* **8**, 25–29.

Thomas, J. F., Anderson, C. E., Downs, R. J. and Raper, C. D. (1973). The effect of carbon dioxide enhancement upon growth of *Nicotiana tabacum*. *Report to the 34th Meeting, Society for Experimental Biology*, Bowling Green, Kentucky.

Warren Wilson, J. (1966). Effect of temperature on net assimilation rate. *Annals of Botany* **30**, 753–761.

Went, F. W. (1957). "Environmental Control of Plant Growth." *Chronica Botanica* **17**, Ronald Press, New York.

Wittwer, S. H. (1967). Carbon dioxide and its role in plant growth. *Proceedings XVII International Horticultural Congress* **3**, 311–322.

Wittwer, S. H. and Robb, W. (1964). Carbon dioxide enrichment of greenhouse atmospheres for food crop production. *Economic Botany* **18**, 34–56.

Wright, S. T. C. (1972). Physiological and biochemical responses to wilting and other stress conditions. *In* "Crop Processes in Controlled Environments" (A. R. Rees *et al.*, eds), pp. 349–363. Academic Press, London and New York.

Zelitch, I. (1971). "Photosynthesis, Photorespiration and Plant Productivity." Academic Press, New York and London.

CHAPTER V

Water and Nutrition

Water

Water supply affects, directly or indirectly, nearly every plant process. It maintains the turgor necessary for plant growth, acts as a solvent for gases and nutrient salts and performs as a reagent in the many hydrolytic reactions that occur in the plant.

Water is a compound of unusual characteristics. It has for example, a very high heat of vaporization, 540 cal g^{-1}, as well as a high surface tension, internal pressure and dielectric constant. A discussion and explanation of the unique properties of water are provided by Slatyer (1967) and Kramer (1969).

LIQUID WATER

Plant growth occurs when water is present in the substrate in quantities between field capacity and the wilting point. Field capacity is defined as the amount of water held by the substrate against the pull of gravity. Although the water tension at field capacity is affected by the structure of the substrate it is usually considered to be about −0·3 bar or about one-third of an atmosphere. One bar is a force equal to 0·987 atm or 10^6 dyn cm^{-2} or 750 mmHg.

Pure water is defined as having an osmotic potential of zero. When the vapor pressure of the system is reduced below that of pure water by the addition of solutes, by lowering the temperature or by capillary forces as are found in soils, the potential is expressed as a negative number. Soil at wilting point has a water potential of approximately -15 bar. Above -0.3 bar the soil becomes saturated and plants that are not adapted to flooding are injured or killed, not because of too much water but because the water displaces the air and the roots are deprived of oxygen. Below -15 bar water will not move from the soil into the roots and plants other than xerophytic ones that contain stored water or are able to go dormant are injured or killed. The water potential can also be lowered by the presence of osmotically active solutes. Therefore, the build-up of salts in a pot can lower the availability of water to the plants.

Soil Moisture. The actual amount of water contained in the substrate between field capacity and wilting point varies with the make-up of the substrate, as does the amount of water remaining in the substrate at the wilting point. Pores smaller than 0.4×10^{-4} mm in diameter will be filled

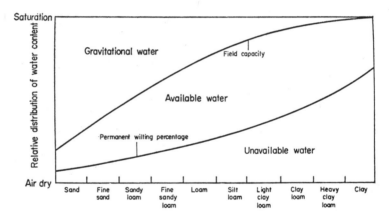

Fig. 1. Soil water availability in relation to texture of soils (Hewlett and Nutter, 1969).

with water at field capacity but only those less than 2×10^{-6} mm will hold water when the soil is at the wilting point. The amount of water held by a given substrate therefore depends upon the number and sizes of the pores. Clay soils, for example, have large numbers of small pores while sand has relatively few of large size. Therefore on a water per volume basis a clay soil at the wilting point can contain more total water than sand does at field capacity (Fig. 1). Obviously, the total amount of water in a soil can be of little interest, the more important factor being the amount of water available for plant growth.

Measuring Soil Moisture. Soil moisture and its loss can be measured by many different methods. Only those more readily applicable to container-grown plants will be considered here. The various methods used to measure the soil moisture content in the field were described by Kramer (1969).

The most direct way to determine substrate moisture in a pot is to determine the water content by weight. This is accomplished by repeatedly weighing the pot with the substrate and plant in it, starting at the time the pot is watered to field capacity. If necessary, a correction can be made at the end of the experiment for plant growth. On a day-to-day basis the error due to plant growth is almost negligible. For larger pots or blocks of substrate the water tension can be measured with a tensiometer. This is a device that consists of a porous ceramic thimble filled with water and attached to a tension gauge. As the substrate dries the water moves through the porous walls of the tensiometer maintaining an equilibrium with the water in the substrate. The water tension is registered on the gauge. An electrical method of measuring substrate moisture involves alterations of electrical resistance between electrodes with changes in moisture. The salt effect is reduced by embedding the electrodes in blocks of plaster of paris. The units are available commercially under the name of Bouyoucos blocks. Their construction and operation are described in a series of papers by Bouyoucos of which the latest was published in 1954.

Psychrometric methods can be used to determine water potential by measuring the depression of the vapor pressure. For this, the rate of water movement from a point source with a known water potential to the substrate is determined. The actual measurement consists of determining the cooling rate of a drop of water that is evaporating from a thermocouple and is being absorbed by the substrate.

Thermocouple psychrometers continue to be improved in design for various applications (Monteith and Owen, 1958; Richards and Otaga, 1958; Rawlins and Dalton, 1967; Hoffman and Splinter 1968). The water potential of the plant can be a good indicator of the water status in the soil, especially as it approaches the wilting point. A small, newly developed, leaf thermocouple psychrometer is described by Hoffman and Rawlins (1972). An indirect measurement of the amount of water available to the plant is the electronic micrometer described by Splinter (1969). The response time of a plant to step inputs of water can be recorded and reductions in daily growth can be detected long before any visible signs of water stress.

The primary water problem in growing plants in chambers and greenhouses has been to maintain the soil water content at a high enough level to avoid a water stress. Water deficiencies frequently develop as the plants grow larger and the researcher may not be aware of the deficiency problem until the plants actually wilt. Some researchers prefer porous clay pots to plastic pots because evaporation of water from the clay pot surface helps to maintain a low soil temperature. However, this is an added source of water loss and when the plants become larger can hasten the onset of wilting or at least restrict growth by reduction of available water.

Maintaining Adequate Soil Moisture. Adequate water supply can be maintained by frequent hand watering or by the use of automatic watering systems. The latter vary from periodic flooding to an almost continuous drip culture. The method used will depend upon the availability of manpower, equipment, substrate media, and the species of plants being grown.

Growing plants in pots, as is done in chambers and greenhouses, poses several unique water problems. The rapid growth of the plant causes it to have a rapidly increasing demand for water. Consequently, the amount of water supplied has to increase and with very large plants an almost steady supply is necessary. Care must be exercised at all times, however, to prevent drowning the roots. Soil, especially if it contains much clay, can be extremely difficult to judge in terms of available water content. Soil blocks neither wet nor dry uniformly and therefore because the surface is wet or dry it does not necessarily mean that the interior or bottom layers have reached the same state. In the two phytotron units of SEPEL this problem is handled by using a potting medium composed of gravel, vermiculite, and in some instances peat. The gravel provides good drainage and root aeration while the vermiculite and peat have a high water-holding capacity. An excess of water is applied to the pots twice a day routinely. For example, 100 ml of excess water is added each time to 11·4 cm pots holding 0·6 l of substrate. Additional watering may be carried out, depending upon the environmental conditions being used and the size of the plants. The water-holding capacity is varied by changing the ratios of the three ingredients to meet the aeration requirements of the particular plants in the study.

A number of automatic watering systems have been used for container-grown plants in growth chambers and greenhouses. Sometimes called autoirrigation, each method has its own particular advantages and disadvantages. One system uses glass wool wicks to move water from a reservoir into the substrate. Another method uses a sand bed kept wet by water from a reservoir. Pots are placed on the sand and water moves up into the substrate by capillary action. Still other systems plunge the pots into a bed of gravel contained in a water-tight bench which is flooded periodically and then slowly gravity drained.

Commercial surface watering systems use individual pot feeders from a main manifold. This system, sometimes referred to as trickle irrigation, can be controlled by timing devices, by tensiometers or by evaporation meters such as described by Christensen (1965). A single tube per container does not work well because the water tends to wet primarily a column of substrate directly beneath the feeder. Christensen (1965) noted that moving the trickle point will retard growth until a new balance can be obtained. More uniform watering is obtained by use of a perforated watering ring on each pot or some system that at least provides several water outlets per pot.

WATER VAPOR

Water in the vapor phase is of primary concern because it influences the rate of transpiration and consequently the withdrawal of water from the substrate. Water vapor is only of importance above the ground. At 15°C the vapor pressure of the substrate atmosphere when it is at the wilting point is less than that of pure water by only 0·25 mmHg. Even at the wilting point at 33°C the substrate atmosphere will have a vapor pressure deficit of less than 1 mmHg which is a relative humidity of 98%.

Relative Humidity. The water content of the air can be measured by several methods. Relative humidity is the parameter that is determined in most instances because of the ease and low cost by which it can be done. One way the relative humidity can be measured is with a wet and dry bulb thermometer, a psychrometer. The evaporation of water when air is passed over the water-soaked cloth-jacketed bulb cools the wet bulb unit: the drier the air passing over the wet bulb the more the temperature is lowered. Using the temperature readings from the wet and dry bulb thermometers the relative humidity can be determined using standard tables that can be found in a physics handbook or from tables that are supplied with the psychrometer.

Relative humidity can also be measured electrically. Instruments are available that use a transducer capable of converting small changes in relative humidity to large changes in electrical resistance by the use of noble metal sensors. They are coated with a highly sensitive hygroscopic film of lithium chloride. The advantage of such an instrument is the rapidity and accuracy with which relative humidity can be measured. Although the sensors do not require the rapid movement of air as do the wet–dry psychrometers, air movement will reduce the response time. There are two main disadvantages to this instrument. One is that each sensor covers a range of less than 20% r.h., so that a series of sensors is needed to cover a more complete range. The second disadvantage is that each unit has to be factory calibrated and the unit has to be recalibrated if it becomes wet, hot or contaminated with any foreign materials.

Another relative humidity sensing device is the hair hygrometer, which operates because hair expands and contracts with changes in water content. These types of sensors, which are relatively slow in their response time and consequently do not detect sudden fluctuations of short duration, are usually found in hygrothermographs and humidistats.

Dew Point. The most accurate and also most expensive method of determining the water content of air is through the use of instruments that measure the dew point. This is the temperature to which an air mass has to be lowered to cause condensation of water. From the dew point reading the vapor pressure can be determined directly from tables for the vapor pressure of water, because the vapor pressure of the air at the dew point is the same as the vapor pressure of pure water at the same temperature (Table I). The vapor pressure of the air is, of course, directly related to the actual water content of the air.

TABLE I

Vapor pressure of pure water and saturated air (mmttg)

T (°C)	VP	T (°C)	VP	T (°C)	VP
−5	3·2	20	17·5	45	71·9
0	4·6	25	23·8	50	92·5
5	6·5	30	31·8	55	118·0
10	9·2	35	42·2	60	149·4
15	12·8	40	55·3	65	187·5

Relative Humidity, Vapor Pressure and Vapor Pressure Deficit. Relative humidity (r.h.) alone is of little value in itself as an indicator of potential water loss from a plant or other evaporative surfaces. Transpiration is basically a water evaporation and vapor gradient problem. Relative humidity is a measure of the amount of water in the air relative to the amount that would be present if the air were saturated at the same temperature. Saturated air and pure water have the same vapor pressure (v.p.) so under these conditions there is no net exchange of water between them.

In Fig. 2 is shown the water content of air in terms of vapor pressure at different relative humidities over a range of temperatures. Note that vapor pressure decreases markedly with a decrease in either temperature or r.h.

Vapor pressure deficit (v.p.d.), however, and not v.p. determines the

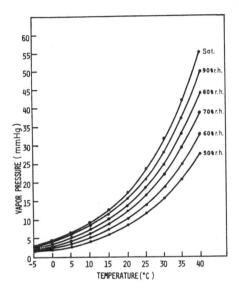

FIG. 2. Vapor pressure of the atmosphere over a range of temperature and degrees of saturation (relative humidities).

rate of evaporation. Thus, the differences between the vapor pressures of the waters in the cells of the leaf, the intracellular spaces and the atmosphere determine the potential rate of evaporation of water from a leaf. The greater the differences the more rapid the rate of water transfer, modified of course by the additional factor of stomatal number and aperture. The v.p.d. increases rapidly with an increase of temperature or with a decrease in r.h. at a given temperature (Fig. 3).

Differences in water stress are often difficult to evaluate in comparing growth chamber results. The reasons for this are that usually only r.h. and not v.p.d. is determined and water content of the atmosphere is difficult and expensive to control, especially if the necessary equipment is not built into the unit.

As a rule of thumb, relative humidity will decrease approximately 4% for each increase of 1°C in temperature provided no water is added to or removed from the atmosphere.

The differences in v.p.d. can and do make large differences in the amount of water used by a plant and thus the rate of water uptake and movement through a plant. To neutralize this effect, chambers should be operated at temperatures and relative humidities that compensate for each other. For example two chambers operating at 80% r.h. but at 15°C and 25°C respectively would have quite different v.p.d.: 2·6 and 4·8 respectively. To operate these chambers at the same v.p.d. while maintaining the 25°C and 80% r.h. in the warmer unit would require that the cooler chamber be kept at 62·5% r.h.

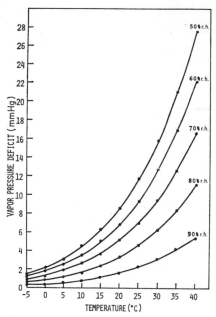

FIG. 3. Vapor pressure deficits at different relative humidities over a range of temperatures.

Vapor pressure has exactly the same relationship with temperature as does water because the vapor pressure of saturated air, 100% r.h., is the same as that for water. Thus the v.p.d. of the air can be determined from the r.h. provided the temperature is known:

$$\text{v.p.} = \text{r.h.} \times \text{s.v.p.}$$

Then, v.p.d. = s.v.p.—v.p.; s.v.p. is the saturated vapor pressure of air at the given temperature and is the same as the v.p. of pure water at the same temperature (Table I).

Cooling chambers and greenhouses through the direct use of water evaporation such as with fan and wet pad systems results in a relatively low v.p.d. in the atmosphere. Conversely, if a direct expansion system involving compressors and a cold coil is used, the v.p.d. of the atmosphere will tend to be large due to the removal of water as it condenses on the coil. This dry air can then cause a rapid rate of water loss from the plants. Some of the new modulating gas control systems or use of secondary coolant with proportioning controls that prevent the coil from fluctuating from extremely hot to extremely cold conditions remove much less water from the air. In both latter systems the coil is maintained at a steady temperature near the chamber set point through the regulated addition of liquidified Freon, cold ethylene glycol, a cold salt solution, or chilled water.

How much drying effect a cold coil will have on plants can be estimated from the air temperature in the chamber and the temperature of the air coming off the coil. The approximation that for each degree centigrade difference the relative humidity will be lowered by 4% requires that the coil is at or below the dew point and the air discharged from the coil is saturated. To keep the dehumidifying effect small the difference between coil temperature and chamber temperature must be kept small. This difference can be kept to a minimum by using a coil which has a large cooling surface with a high heat exchange capacity. The large cooling surface permits the coil to be operated at a temperature not much below the desired air temperature. The heat transfer efficiency of the coil also can be increased by keeping the coil wet.

If the coil is undersized the high velocity of the air does not allow sufficient time for the air to come to temperature equilibrium with the coil. Consequently, the coil has to be operated at a lowered temperature to obtain the needed cooling. If the coil has to be operated below 0°C for long periods it will become covered with ice that restricts the air flow. Another problem with high velocity air over a coil is that the air will entrain condensate from the coil and carry liquid water into the system.

ICE

Water in the solid state is ice and usually is of little concern unless a chamber goes out of control or the experiment calls for subfreezing conditions. Water in the soil in ice form is neither available nor usable by plants. Water, when it freezes, expands in volume about 10%. Ice can and usually does cause more physical than physiological damage, especially within the plant itself where the ice crystals rupture cell membranes. When a plant and the air are at or below the freezing point the water requirements of the plant are low due to the low v.p.d. potential of the air. At 0°C saturated air only has a v.p. of 4·6 mmHg (Table I). However, if the plant or soil water is frozen and the air temperature is raised rapidly the vapor pressure gradient between the air and plant rises also. Under these conditions the plant may be killed by desiccation due to the immobility of water in the ice form in the plant and soil. Extremely cool or cold conditions are sometimes required for studies. Such work often requires special attention to equipment, especially the refrigeration and air handling systems. Tundra type environments including permafrost levels have been used in chambers of the Duke University phytotron to study plant growth and development under extreme conditions (Trent, 1972).

Nutrition

Nutrients as used here will refer to those elements that are essential for plant growth and which are primarily taken into the plant through the root system. Hydrogen enters the plant as part of the water molecule and is discussed above. The two elements that enter as gases—carbon in the form of carbon dioxide and oxygen as O_2—are discussed in Chapter IV.

Plants grown in controlled-environment facilities sometimes develop abnormalities and difficulties not usually apparent in field or greenhouse practices. While light, temperature or carbon dioxide levels may be the cause of some problems, many can be traced at least in part to nutrition. Went (1957) pointed out that "not many experiments have been carried out on nutrition of plants under different temperature and light conditions". This statement continues to be true today.

NUTRITIONAL DEFICIENCIES

Plants growing slowly in soil may not take up the nutrients faster than they are released from the soil through normal degradation processes. However, if the growth rate is increased by having the plants under optimum environmental conditions the uptake rate may exceed the natural soil release rate. Thus, nutrient deficiencies frequently develop in plants grown in soil in chambers under near optimum temperature, light and water conditions. Under such conditions the total amount of nutrients in the soil or at least the pool of available nutrients is rapidly depleted.

Nutritional deficiency symptoms are quite specific and depend upon the element that is deficient and the plant species (Sprague, 1964; Epstein, 1972; Gauch, 1972). Some symptoms are rather general regardless of the plant species. The yellow color of leaves due to a nitrogen deficiency and the purple color caused by a phosphorus deficiency are typical.

CULTURAL TECHNIQUES AND MEDIA

The prevention of nutritional deficiencies in plants growing under conditions that promote rapid growth requires that the plants have a readily available supply of nutrients. If the plants are growing in pots of soil, fertilizer has to be applied more frequently than in the field or in the greenhouse. The elements of nitrogen, phosphorus, and potassium, which are used in large quantities by plants, are usually the ones that need to be added frequently. In addition, the soil in a pot may need to be changed periodically to assure a satisfactory balance of nutrients.

For much research work and in some commercial greenhouse applications, soil lots obtained over a period of time are much too inconsistent. Nutrient cultures have proven much more reliable than soil in these cases because the nutrient environment is under the investigator's control. Uniformity of plant growth from one set of studies to the next is much greater since the nutritional system remains the same. The nutrition can be altered at any time so that nutritional sequences can be matched to the various stages of plant development or adjusted as a function of the environment.

Water culture where the roots are submerged in a nutrient solution works well, but requires considerable aeration and is generally more clumsy to use than the nutrient culture using an inert substrate. The most common substrates used in nutrient cultures are sand, gravel, perlite, vermiculite, peat or various combinations of these materials. The Duke University phytotron uses a mixture of gravel and vermiculite in much the same manner as the Earhart Laboratory. The NCSU phytotron uses gravel and a commercial peat–lite mix as a standard substrate but gravel and vermiculite and various kinds of sand are also used. The French phytotron uses a glass wool substrate (Nitsch, 1972). Nutrient solution is generally applied by the "slop" culture technique where enough excess nutrient is applied to the pot to allow a substantial flushing action. Drip or trickle methods of automatic watering can also be used, as can the various methods of automatic irrigation.

There is no nutrient solution that is optimum for all plants, or for that matter for all stages of growth of a single kind of plant. In greenhouse practice different mixtures are required to provide the best nutrient solution for plants grown at different times during the year. However, equally good plant growth occurs in soils that have very different compositions, and equally good growth can be obtained over a wide range of nutrient solution combinations. Thus the composition of plant culture solutions has remained basically the same since Knops and Sachs first used them in 1860 (Sachs, 1887). The modified "Hoagland solution" is probably the most frequently applied nutrient solution used in the United States today (Hoagland, 1944; Hoagland and Arnon, 1950). It differs very little from solutions used in Europe. Slight changes have been made involving the ratios of the various nutrients and new forms of nutrient chemicals have been incorporated. These latter changes include using chelated iron and other minor elements to increase their availability over a wider range of pH. A detailed description of culture solutions and their use is to be found in Epstein (1972).

Various laboratories that use large quantities of nutrient solution have made modifications in the solution to fit specific needs. A comparison of several such solutions is presented in Tables II–VII. Most laboratories

TABLE II

Composition of the Hoagland nutrition solution used in the Earhart Laboratory

Stock solution	Molecular weight	g l^{-1} in stock solution
(1) Calcium nitrate $Ca(NO_3)_2$	164·1	364·0
(2) Potassium nitrate KNO_3	101·1	221·28
(3) Magnesium sulfate $MgSO_4.7H_2O$	120·4	217·6
(4) Potassium dihydro phosphate KH_2PO_4	136·1	62·08
(5) Mixed micronutrients, containing		
$CuSO_4.5H_2O$	249·7	0·0354
$MnSO_4$	151·0	0·609
$ZnSO_4$	161·4	0·0974
H_3BO_3	61·8	1·269
$H_2MoO_4.4H_2O$	234·0	0·0398
(6) Fe–Sequestrene, prepared by mixing		
KOH		6·8
Sequestrene AA		10·4
$FeSO_4.7H_2O$		10·0

Diluted for use at rate of 1 l of each of the six stock solutions plus 400 l water (from Went, 1957)

make up stock solutions that are diluted, often automatically with some kind of proportioning pump, before being mixed. If the concentrated stock solutions are mixed before dilution some of the essential elements precipitate as insoluble salts. Between 1957 and 1965 the solution used in the Earhart laboratory was modified and the number of stock solutions was reduced from six to two (Tables II and III). Most automatically diluted nutrient systems now use only two stock solutions.

The mineral concentration of culture solutions is many times greater than that found in soil solutions. This high mineral concentration is used primarily because it is difficult to maintain a low concentration against the withdrawal by the plant roots. An important quality of soil is its exchange capacity or ability to bind and then release nutrients to maintain their availability. The nutrients in the soil solution are thus being continually

TABLE III

One-half strength modified Hoagland as used in Earhart-Campbell

Name	g l^{-1} stock solution	ppm working solution	
Calcium nitrate	295	Ca	103·0
$(Ca(NO_3)_2.4H_2O)$	(53100)	N	70·0
Sequestrene	38·44	Fe	10·0
[NaFe (13%)]	(6920)		
Potassium phosphate	34·25	K	19·63
(KH_2PO_4)	(6165)	P	15·53
Potassium nitrate	126·65	K	98·0
(KNO_3)	(22800)	N	35·1
Magnesium sulfate	126·65	Mg	24·9
$(MgSO_4.7H_2O)$	(22800)	S	32·9
Zinc sulfate	0·0555	Zn	0·0253
$(ZnSO_4.7H_2O)$	(10)	S	0·0124
Manganous sulfate	0·3905	Mn	0·254
$(MnSO_4.H_2O)$	(70·3)	S	0·148
Copper sulfate	0·0206	Cu	0·0105
$(CuSO_4.5H_2O)$	(3·71)	S	0·0053
Boric acid	0·725	B	0·254
(H_3BO_3)	(130·5)		
Molybdic acid	0·00463	Mo	0·0052
$(MoO_3.2H_2O)$	(0·833)		
$(MoO_3.Anhydrous)$	0·0037		
	(0·667)		

1. Figures in parentheses are quantities required for 180 l stock solution.

2. Calcium nitrate and sequestrene are combined in solution "A"; all other salts are in solution "B"

3. Both stock solutions "A" and "B" are to be used at 1:500 or 200 ml stock per 100 l water

TABLE IV

A nutrient solution that includes chlorine
(Johnson et al., 1957). Modified by Epstein (1972)[c]

Macronutrients

Compound	Molecular weight	Concentration of stock solution (M)	Concentration of stock solution g/l	Volume of stock solution per litre of final solution (ml)	Element	Final concentration of element (μM)	Final concentration of element (ppm)
KNO_3	101·10	1·00	101·10	6·0	N	16000	224
$Ca(NO_3)_2 \cdot 4H_2O$	236·16	1·00	236·16	4·0	K	6000	235
$NH_4H_2PO_4$	115·08	1·00	115·08	2·0	Ca	4000	160
$MgSO_4 \cdot 7H_2O$	246·49	1·00	246·49	1·0	P	2000	62
					S	1000	32
					Mg	1000	24

Micronutrients

Compound[a]	Molecular weight	Concentration of stock solution (mM)	Concentration of stock solution g/l	Volume of stock solution per litre of final solution (ml)	Element	Final concentration of element (μM)	Final concentration of element (ppm)
KCl	74·55	50	3·728		Cl	50	1·77
H_3BO_3	61·84	25	1·546		B	25	0·27
$MnSO_4.H_2O$	169·01	2·0	0·338	1·0	Mn	2·0	0·11
$ZnSO_4.7H_2O$	287·55	2·0	0·575	1·0	Zn	2·0	0·131
$CuSO_4.5H_2O$	249·71	0·5	0·125		Cu	0·5	0·032
$H_2MoO_4(85\% MoO_3)$	161·97	0·5	0·081		Mo	0·5	0·05
Fe–EDTA[b]	346·08	20	6·922	1·0	Fe	20	1·12

[a] A combined stock solution is made up containing all micronutrients except iron

[b] Ferrous dihydrogen ethylenediamine tetraacetic acid

[c] Modified to substitute chelated iron for the iron tartrate in the original formulation

TABLE V

Composition of the standard nutrient solution used in the Gif phytotron
for all plant species (Nitsch, 1972)

In mg litre^{-1}		*In* μmol litre^{-1}	
KNO_3	411	NO_3^-	12·134
$Ca(NO_3)_2.4H_2O$	959	SO_4^{--}	3·272
$(NH_4)_2SO_4$	137	PO_4^{---}	1·007
$MgSO_4.7H_2O$	548	$EDTA^{----}$	110
KH_2PO_4	137	BO_3^{---}	50
$EDTA\ Na_2Fe.2H_2O$	41	Cl^-	36
H_3BO_3	3	MoO_4^{--}	0·2
KCl	2·7	K^+	5·111
$MnSO_4.H_2O$	1·7	Ca^{++}	4·063
$ZnSO_4.7H_2O$	0·27	Mg^{++}	2·223
$(NH_4)_6Mo_7O_{24}.4H_2O$	0·27	NH_4^+	2·077
$CuSO_4.5H_2O$	0·13	Na^+	110
		Fe^{+++}	110
		Mn^{++}	10
		Zn^{++}	0·9
		Cu^{++}	0·5

In practice, the solution is prepared as follows: (1) the tank is
filled with 3650 litres of deionized water; (2) the four stock
solutions are added in the order given with constant stirring:

Solution A: KNO_3 (1 kg) + KH_2PO_4 (0·5 kg) in 10 l of H_2O

Solution B: $MgSO_4.7H_2O$ (1 kg) + $(NH_4)_2\ SO_4$ (0·5 kg) in
10 l of H_2O

Solution C: $Ca(NO_3)_2.4H_2O$ (4 kg) + EDTA Na_2Fe (0·15
kg) in 10 l of H_2O

Minor elements: KCl (10 g) + H_3BO_3 (11 g) + $MnSO_4.H_2O$
(6·2 g) + $ZnSO_4.7H_2O$ (1 g) + $(NH_4)_6Mo_7O_{24}.4H_2O$
(1 g) + $CuSO_4.5H_2O$ (0·5 g) + H_2SO_4 (0·5 ml) in 1 litre of
H_2O

TABLE VI

North Carolina State University phytotron nutrient solution

Name	Stock solution (g l⁻¹)	Working solution (ppm)
	Solution "A"	
Magnesium nitrate	65·0	Mg 12·32
		N 14·20
Calcium nitrate	160·0	Ca 54·31
		N 37·96
Sequestrene 330 Fe	25·0	Fe 5·00
	Solution "B"	
Ammonium nitrate	80·0	N 55·99
Potassium phosphate:		
monobasic	12·0	K 6·89
		P 5·46
dibasic	14·0	K 6·28
		P 4·98
Potassium sulfate	15·0	K 13·46
		S 5·52
Sodium sulfate	17·0	Na 11·01
		S 7·67
Boric acid	0·700	B 0·24
Molybdic acid	0·005	Mo 0·005
Hampene Zinc	0·045	—
Hampol Manganese	0·630	—
Hampol Copper	0·030	—
Sequestrene Cobalt	0·001	—

1. Stock solutions are proportioned at the rate of 1 ml of A and 1 ml of B to 500 ml of water.
2. Uranine (sodium fluorescein) is added to the stock "B" at 0·25 g l⁻¹ of stock. This gives the working solution a green colour so it can be distinguished from water during application.

TABLE VII

Nutrient solution used at the Duke phytotron (approximately one-half strength Hoaglands)

Chemical	$g\ l^{-1}$ stock solution	Working solution
"A" solution		ppm
(1) Calcium nitrate	236·2	Ca 80·0
		N 56·0
"B" solution		
(2) Potassium phosphate	34·25	K 19·63
		P 15·53
(3) Potassium nitrate	126·65	K 98·0
		N 35·1
(4) Magnesium sulphate	126·65	Mg 24·9
		S 32·9
(5) Sequestrene (330 Fe DTPA)	38·44	Fe 10·0
(6) Zinc sulfate	0·0555	Zn 0·0253
		S 0·0124
(7) Manganous sulfate	0·3905	Mn 0·254
		S 0·148
(8) Copper sulfate	0·0206	Cu 0·0105
		S 0·0053
(9) Boric acid	0·725	B 0·254
(10) Molybdic acid	0·00463	Mo 0·0052
(11) (MoO_3—anhydrous)	0·0037	—
(12) Ammonium nitrate	50·0	N 35·0
(13) Ammonium phosphate	29·0	P 15·0
		N 7·0
(14) Sodium chloride	5·85	Na 4·6
		Cl 7·1

1. Add to 500 parts water; 1 part "A" and 1 part "B" stock solution.
2. Items 2 through 11 are a commercially prepared mix

replaced by the minerals absorbed on the soil particles; a process that does not occur in nutrient culture.

If a planting medium other than soil is used the frequency with which the nutrient solution is to be added needs to be considered. Pure sand will require the solution to be added very frequently and the frequency will depend upon the ratio of container size to the size of the plant. Sand has a very low water-holding capacity and few binding sites; therefore, it has a low exchange capacity for minerals. Vermiculite, exploded mica, is frequently mixed with sand or gravel to form a potting medium. The vermiculite has a high water-holding capacity and a high exchange capacity.

This decreases the frequency with which the nutrient solution needs to be supplied to the container. Peat also has a high water-holding and ion exchange capacity and may be added to or used in place of vermiculite. The water-holding, drainage and aeration characteristics of the media can be varied to satisfy the requirements of different plant species by changing the ratios of sand to vermiculite and peat.

Plants absorb minerals differentially. Nutrient solutions which at the start are usually on the acid side (pH 5–7) become alkaline. The shift to alkalinity tends to put several of the elements into a non-soluble form and therefore are no longer available to the plants. Iron and phosphorus are the two most severely affected. Also, the differential removal of ions by plants causes an imbalance in the nutrient solution and build-up of salts. In the two-unit phytotron of SEPEL to prevent an unbalancing of the nutrients the pots are flushed with a nutrient solution daily. This, at least theoretically, returns the retained nutrients to a proper balance by mass action of the salts. In practice the results, judged by plant growth, show this to be a satisfactory way to operate. However, both on a water supply basis and a nutrient supply basis there are indicators that a drip or trickle culture might be even better. This procedure involves very frequent applications of water and nutrients during the day and would require an automatic operating system.

NUTRITIONAL REQUIREMENTS OF PLANTS

Ingestad (1971) has conducted extensive research on nutritional requirements of birch seedlings. He was able to obtain maximum growth, not by changing the culture solution frequently as is usually done, but rather by

TABLE VIII

Optimum nutrient proportions with nitrogen set at 100 for birch seedlings

Macronutrients		Micronutrients	
N	100	Fe	0·7
K	65	Mn	0·4
P	13	B	0·2
S	9	Cu	0·03
Ca	7	Zn	0·03
Mg	8·5	Cl	0·03
		Mo	0·007
		Na	0·003

(Modified from Ingestad, 1971)

analyzing the solution and adding only that which had been removed. The chemical proportions in his starting solution corresponded to that within the plant (Table VIII). For pine and spruce, on the basis of nitrogen at 100, the proportion of potassium is only 50 compared to the 65 for birch. For vegetables the proportion can go as high as 90. (These latter two figures were obtained through personal correspondence). While there are differences in requirements for plants of different species the soil solution in fertilized soils, and especially nutrient solutions where they are used, usually contain more than sufficient amounts to mask these differences in demands.

It is not our intention to delve into the details of nutrition requirements because this has been done by a number of authorities on the subject (Hoagland, 1944; Sprague, 1964; Epstein, 1972; Gauch, 1972). Some consideration, however, should be given to apparent changes in nutritional requirements or the ability of the plant to absorb and translocate certain elements that occur as a function of environment, even though very little detailed information seems to be available.

As temperatures are lowered a point is reached where plants become chlorotic, irrespective of the applied nutritional level. For example new leaves of Okra plants grown at 18/14°C day/night temperatures emerge almost devoid of chlorophyll. As the leaves expand chlorophyll will begin to appear and by the time they are fully expanded chlorophyll may be more or less uniformily distributed throughout the leaf. Marigolds, such as the variety Bolero, develop a good deal of anthocyanin at 18/14°C and are quite chlorotic at 18/22°C day/night temperatures. At 22/18°C however, no nutritional difficulties are apparent. Nitrate is the ion most affected by temperature and Williams and Vlamis (1962) report that almost none is absorbed at 13°C.

Tobacco which is chlorotic at 18/14 or below is dark green at 34/30 day/night temperatures but the leaf margins are inrolled and the lamina is savoyed; a description similar to the one often used for Ca deficiencies. A similar finding was reported by Chang et al. (1968). Corn was also noted to produce Ca deficiency symptoms as temperatures increased to about 35°C (Walker, 1969).

Deficiency symptoms of one kind or another may also appear as the level of radiant energy is raised or at the high light intensities of the modern plant growth room, as the duration of the high intensity light period is increased.

CO_2 levels several times greater or lower than the normal 350 ppm can induce severe chlorosis that may or may not be alleviated by a change in the nutrient solution and high humidity is reported to cause Ca deficiency in cauliflower (Wiebe and Krug, 1973).

References

Bouyoucos, G. J. (1954). New type electrode for plaster of paris moisture blocks. *Soil Science* **78**, 339–342.

Chang, S. Y., Lowe, R. H. and Hiatt, A. J. (1968). Relationship of temperature to the development of calcium deficiency symptoms in *Nicotiana tabacum*. *Agronomy Journal* **60**, 435–436.

Christensen, S. A. (1965). Irrigation and its automation in glasshouses. *Acta Horticulturae* **2**, 76–80.

Epstein, E. (1972). "Mineral Nutrition of Plants: Principles and Perspectives." John Wiley, New York.

Gauch, H. G. (1972). "Inorganic Plant Nutrition." Dowden, Hutchinson and Ross, Stroudsburg, Pa.

Hewlett, J. D. and Nutter, W. L. (1969). "An Outline of Forest Hydrology." *University of Georgia Press, Athens, Ga.*

Hoagland, D. R. (1948). "Lectures on the Inorganic Nutrition of Plants." Chronica Botanica, Waltham, Mass.

Hoagland, D. R. and Arnon, D. I. (1950). The water-culture method for growing plants without soil. *California Agricultural Experiment Station*, Circular 347.

Hoffman, G. J. and Rawlins, S. L. (1972). Silver-foil psychrometer for measuring leaf potential *in situ*. *Science* **177**, 802–804.

Hoffman, G. J. and Splinter, W. E. (1968). Instrumentation for measuring water potentials of an intact plant-soil system. *Transactions of the American Society for Agricultural Engineering* **11**, 38–40.

Ingestad, T. (1971). A definition of optimum nutrient requirements in birch seedlings. *Physiologia Plantarum* **24**, 118–125.

Johnson, C. M., Stout, P. R., Broyer, T. C., and Carlton, A. B. (1957). Comparative chlorine requirements of different plant species. *Plant and Soil* **8**, 337–353.

Kramer, P. J. (1969). "Plant and Soil Water Relationships." McGraw Hill, New York.

Monteith, J. L. and Owen, P. C. (1958). A thermocouple method for measuring relative humidity in the range of 95–100%. *Journal of Scientific Instruments* **35**, 443–446.

Nitsch, J. P. (1972). Phytotrons: Past achievements and future needs. *In* "Crop Processes in Controlled Environments", pp. 33–35. (A. R. Rees, K. E. Cockshull, D. W. Hand and R. G. Hurd, eds). Academic Press, London and New York.

Rawlins, S. L. and Dalton, F. N. (1967). Psychrometric measurement of soil water potential without precise temperature control. *Proceedings Soil Science Society of America* **31**, 297–301.

Richards, L. A. and Otaga, G. (1958). Thermocouple for vapor pressure measurement in biological and soil systems of high humidity. *Science* **128**, 1089–1090.

Sachs, J. von. (1887). "Lectures in the Physiology of Plants." Clarendon Press Oxford.

Slatyer, R. O. (1967). "Plant–Water Relationships." Academic Press, London and New York.

Splinter, W. E. (1969). Electronic micrometer continuously monitors plant stem diameter. *Agricultural Engineering* **50**, 220–221.

Sprague, H. G. (1964). "Hunger Signs in Crops; A Symposium" (3rd edition). McKay Co., New York.

Trent, A. (1972). "Measurement of root growth and respiration in arctic plants." M. S. Thesis. Duke University, Durham, North Carolina.

Walker, J. M. (1969). One-degree increments in soil temperatures affect maize seedling behavior. *Proceedings Soil Science Society of America* **33**, 729–736.

Went, F. W. (1957). "The Experimental Control of Plant Growth." Chronica Botanica, Waltham, Mass.

Wiebe, H. J. and Krug, H. (1973). Physiological problems of experiments in growth chambers. *In* "Basic problems of protected vegetable cultivation'" *International Society for Horticultural Science Symposium*, Hannover, Ger. (in press).

Williams, D. E. and Vlamis, J. (1962). Differential cation and anion absorption as affected by climate. *Plant Physiology* **37**, 198–202.

CHAPTER VI

The Environmental Complex

The "Normal" Plant

Many plant scientists have expressed doubts that "normal" or "natural" plants can be grown under the completely artificial conditions of the controlled-environment room. The same question is far less frequently raised about greenhouse-grown plants although the environmental conditions are at least equally unnatural as they are in the plant growth chamber.

What exactly is this "normal" plant so often alluded to by the biologist? A search of the literature and discussions with colleagues reveal that the normal plant can rarely be defined with any degree of detail. Instead it is usually referred to as: "like plants grown under natural conditions". Unfortunately, "natural conditions" is too broad a term to aid greatly in defining the normal plant. For example, pine seedlings grown in the nursery bed and transplanted to logged areas in reforestation programs do not have the same degree of normality as those that arise in the understory from seed disseminated directly from the trees. Both types of seedlings have been grown under natural conditions but some sets of conditions are more natural than others.

All plants of a species are not alike when grown under natural conditions. Ecophenes which differ in habit, vigor, branching, leaf size and shape and undoubtedly chemical constituents occur as a result of environmental influences. Physiological ecotypes based on the geographical area of the seed source are well documented for woody plants (Pauley and Perry, 1954; Vaartaja, 1954; Wakeley, 1954; Downs and Piringer, 1958). Soybean varieties classified into groups with similar latitudinal responses are in essence photoperiodic ecotypes. The point, therefore, is that plants such as *Pinus*

117

ponderosa grown from seed originating in British Columbia are not the same as the *P. ponderosa* plants grown from Mexican seeds, yet they are all "normal" plants of the same species.

Plants like those in the field or forest can be grown under controlled-environment conditions; within the limits of our ability to define the normal plant. Tobacco is one of the few plants that can be described rather precisely. A 3 year average of Coker 319, for example, reveals that the yield should be about 2480 kg ha^{-1}. This variety would flower in 55 days and be 1·19 m tall with 21 leaves per plant. The leaves would be about 28 × 62 cm at the tenth node. The cured leaf would contain 3·21% nicotine, 0·08% nornicotine, 12·69% reducing sugars and 2·41% total nitrogen (Rice *et al.*, 1970). At several sites in a given year some variation occurs due to moisture conditions and temperatures, as well as soil types and differences in cultural practices. Thus, days to flower may range from 53 to 61, the number of leaves from 19 to 24 and the height from 1·12 to 1·30 m.

The normal field nutritional schedule poses an additional strain on the skills of the investigator to produce a field-grown type tobacco plant in controlled environments. An incorrect nutrition schedule in either the field or the controlled environment can cause abnormally high nitrogen in the plant tissue, delayed leaf ripening, an imbalance in the chemical constituents of the leaf and in the end result in leaves that fail to cure properly (Raper, 1971; Raper and Johnson, 1971). Moreover, early environment can alter and in fact predetermine later growth characteristics such as leaf shape (Raper, 1972; Raper and Thomas, 1972), and for proper growth in chambers and greenhouses a seasonal progression of temperature must be simulated.

One of the many advantages of controlled-environment facilities is the potential for growing any kind of plant necessary for study: abnormal ones or ones as normal as the investigator can describe.

Natural Conditions *vs* Controlled Environments

Another term frequently introduced by plant scientists using controlled-environment facilities is "natural conditions". Questions are raised about the ability to program a plant growth chamber to provide outdoor conditions. Such programming can be done, but why spend huge sums of money on controlled-environment facilities to obtain what is already available: namely, fluctuating environmental factors? The main purpose of phytotrons is to avoid such uncontrolled conditions so that the effects of the various components of the environment can be studied in detail.

Attempts to keep things more natural has often led to complete abandonment of scientific principles. For example, many investigators insist that 12 h of high intensity light must be used because it is more natural than say a 10 h light period. In North Carolina at least, neither daylength appears during the growing season, so the premise for the 12 h period is untrue. Programming the light to simulate sunrise to sunset has not proven to have any advantage over constant irradiance (Hughes and Cockshull, 1971; Raper *et al.*, 1973) although Hoare (1972) stated that the square wave day/night light program causes plants to go into oscillations of rapid water loss. Moreover, a single program rarely occurs in nature more than once during a growing season. Thus a single diurnal program used throughout an experiment is no more natural than a steady daily irradiance.

Daily temperature programs seem to have little value. A day/night temperature regime can satisfy any thermoperiod requirements with less difficulty than programmed temperatures. Moreover, where controlled-environment rooms have been equipped with temperature programmers the majority use only one program which might be based on the average maximum–minimum daily temperatures, or it might be simply a program that appeals to the investigator. Admittedly, such crystal ball methods are used because of the almost complete lack of data upon which to base daily programs of environmental factors, but the result cannot be considered any more natural than on/off, day/night systems. Daily programs have no real resemblance to what occurs out-of-doors and repeatedly using the same program is unrealistic since nature rarely if ever exactly repeats herself even on a daily basis.

Seasonal programming of environmental factors, especially temperature, humidity, water and nutrition, is far more important to biological investigations than daily programs. Simulating natural temperature and nutritional progressions can be done in steps as with tobacco (Raper, 1971; Raper and Johnson, 1971), or a series of optimum temperatures for each stage of growth can be developed. With cotton, for example, optimum germination occurs at 33°C but after 3 days the temperature should be lowered to 27°C (Carns and Mauney, 1968). Since bolls are usually the objective, the plants should be grown at 25/20°C for earliest initiation and normal fruit set. Maximum boll weights are produced at 27/22°C (Hesketh and Low, 1968). Faster boll maturity was obtained at higher temperatures although they were considerably smaller.

Many plant scientists exhibit concern about natural light; not just solar intensities but also solar spectral distributions, presumably including skylight. The reasons cited for this concern are usually vague and are based on the idea that only what is natural is valid. A great deal of research needs to be done but with our current state of knowledge there seems to be no evidence that daily peak solar intensities are necessary for good plant growth if shading is minimized. The requirements for plant growth seem to be adequate light for the photosynthetic system to function and produce an excess of product in the leaves, phytochrome in the proper ratio of forms, and the accessory pigments activated. Artificial light can do these things as well as sunlight without an exact matching of the spectral distribution curves.

There are, of course, a number of special areas where at least partly matching spectral irradiances would be of particular merit. For example, studies with high-altitude vegetation may require increased ultraviolet and plants that occur as understory species in ecological communities or in multiple cropping may need adjusted light quality if "normal" plants are to be developed.

A clear concept of the environmental complex and its relation to plant growth is hampered by lack of knowledge and clouded by our acceptance of climatological information without considering what it means in terms of plant physiology. Solar radiation may be taken from tables as total energy per day. However, these values may be physiologically unrealistic without information on plant response. Peak illuminance is even more illogical because the sun provides such illumination on only some of the days and even then for only a brief period. Average temperature is of little value because extreme highs and lows may occur frequently enough to have a marked effect on the biological system. To look at an extreme but all too frequently occurring example, a daily average of 15°C on a spring day is of little consolation to the grower if it includes an early morning −5°C.

Translation between Field and Laboratory

In the phytotron the environmental complex can be more clearly defined than it can in the field. For a fixed irradiance and CO_2 concentration the optimum day and night temperatures can be obtained. These optima can then be re-evaluated as a function of total energy. The influence of water stress, nutrition, soil aeration and other parameters of the rhizosphere on the optimum aerial environment can easily be determined. Nevertheless, there is a general lack of faith in our ability to transpose phytotron results to plants growing in the field. Actually, the problem is not real; it simply has received little attention, partly because of lack of funds and partly because of the low glamour index of such studies. It seems quite clear that controlled-environment studies fail to coordinate with field results because in most instances the techniques are incorrectly applied. Some mathematical models based on such studies are showing promising results such as "Simcot" (Baker et al., 1972) while others fail and will continue to do so until the approach to the problem is changed through a better under-standing of the physiological processes.

In the phytotron, flue-cured type tobacco plants flower rapidly but fail to set seed at 35/30°C day/night temperatures. Very likely this is a result of high night temperature, if Went's hints (1957) are correct. However, for field interpretation we need to know how many 30°C hours per night are necessary to inhibit seed set and whether there is a critically sensitive period during flower development.

Went (1957) and Haroon et al. (1972) noted that low night temperatures altered tobacco leaf shape and made them more elongate. Leaf shape changes have long been noticed in the field and early phytotron studies were quite inadequate in explaining the phenomenon. Later studies by Raper and Thomas (1972) showed that the temperature during the first 10 days after transplanting determined the final shape of the leaves. Day/night temperatures of 18/14°C produced the lowest leaf length/width ratio. A 4°C *rise* of either the day or night temperature increased the elongate shape. Although an increase in both day and night temperatures gave the largest elongation, a 4°C day increase caused a greater elongation than a similar increase at night.

Note that these data (Raper and Thomas, 1972) do not agree with the conclusions of Went (1957) and Haroon et al. (1972). Also note that the latter investigators used the same day/night temperature throughout the growth of the plant: a common practice in controlled-environment facilities, but not in agreement with the seasonal temperature progression of the natural environment. Most important is that the Raper and Thomas (1972) data were later verified by field observations (Raper, 1972).

Use of controlled-environment facilities to contribute to ecological or agricultural knowledge requires that information from the phytotron relate to field problems and plant behavior. This means that the dynamic nature of the environmental complex must be kept in mind at all times while designing phytotron experiments. Time aspects of environmental factor variations are very important considerations and the extreme values may have considerable influence on plant behavior even when they are not extreme enough to be considered a stress condition. Any factor that differs significantly from those in the natural habitat could affect the correlation between phytotron and field research. The key word here is "significantly". Since not enough research has been done to delineate the allowable differences, investigators either ignore the whole business and keep the environment constant, or they attempt to program all changes, usually on a diurnal basis. Raper's work indicates that natural progressions and variations can be simulated in rather broad steps with very realistic results.

Use of a Fixed and Reproducible Environment

It is not imperative that results of controlled environment work be related to the field in order for the research to be worthwhile. Current techniques of using controlled environment facilities have proven quite adequate for producing constant plant materials and for the many studies aimed at enlarging our knowledge of how plants grow.

Many studies require plants to be grown up to a certain stage of development prior to beginning experimental treatments. Since the plants should be exactly alike at the start of each study they must be grown under a standard, constant set of environmental conditions. The effects of environmental stress, chemical applications and other imposed conditions can be evaluated and various plant functions such as transpiration and photosynthesis can be measured on relatively uniform plant material, resulting in a minimum of background fluctuation. For example, when alternariol monomethyl ether (AME), a substance produced by the leaf spot fungus *Alternaria tenuis*, was tested for its ability to induce chlorosis in tobacco leaves very inconsistent results were obtained with plants grown in the fluctuating environments of ordinary greenhouses. However, standard plants grown in controlled, reproducible environments responded uniformly and consistently to injections of AME (Pero and Main, 1970).

Our progress towards an understanding of the mechanisms of photoperiodism and photomorphogenesis would surely have been severely retarded without controlled-environment facilities, yet very few of these studies attempted to relate to field or natural conditions. It was enough simply to know and understand a little more about how such systems operated; the applications would take care of themselves.

The effects of the interaction of environmental conditions on plant growth are not well understood. At present, progress can be made mainly by investigating the effects of one environmental factor at a time with all others held at constant levels. It should be stressed that the effects of individual environmental factors should be understood before trying to study the interactions. Data from a multifactor experiment conducted in a controlled environment are as impossible to untangle and understand as data collected in an uncontrolled environment. When the effects of individual factors are understood then the levels of various factors can be altered simultaneously and the response of the plant evaluated.

References

Baker, D. N., Hesketh, J. D. and Duncan, W. G. (1972). Simulation of growth and yield in cotton. I. Gross photosynthesis, respiration, and growth. *Crop Science* **12**, 431–435.

Carns, H. R. and Mauney, J. R. (1968). Physiology of the cotton plant. *In* "Cotton: Principles and Practices" (F. C. Elliott, M. Hoover, and V. K. Porter, eds), pp. 43–63. Iowa State University Press, Ames, Iowa.

Downs, R. J. and Piringer, A. A. (1958). Effects of photoperiod and kind of supplemental light on vegetative growth of pines. *Forest Science* **4**, 185–195.

Haroon, M., Long, R. C. and Weybrew, J. A. (1972). Effect of day/night temperatures on factors associated with growth of *Nicotiana tabacum* in controlled environments. *Agronomy Journal* **64**, 509–515.

Hesketh, J. D. and Low, A. (1968). Effect of temperature on components of yield and fiber quality of cotton varieties of diverse origin. *Cotton Growing Review* **45**, 91–100.

Hoare, E. R. (1972). Remarks in the discussion section. *In* "Phytotronique II." (P. Chouard and N. de Bilderling, eds), pp. 368–370. Gauthier-Villars, Paris.

Hughes, A. P. and Cockshull, K. E. (1971). A comparison of the effects of diurnal variation in light intensity on growth of *Chrysanthemum morifolium*. *Annals of Botany* **35**, 927–932.

Pauley, S. S. and Perry, T. O. (1954). Ecotypic variation of the photoperiodic response in Populus. *Arnold Arboretum* **35**, 168–188.

Pero, R. W. and Main, C. E. (1970). Chlorosis of tobacco induced by alternariol monomethyl ether produced by *Alternaria tenuis*. *Phytopathology* **60**, 1570–1573.

Raper, C. D. (1971). Factors affecting the development of flue-cured tobacco grown in artificial environments. III. Morphological behavior of leaves in simulated temperature, light duration and nutritional progressions during growth. *Agronomy Journal* **63**, 848–852.

Raper, C. D. (1972). Temperatures in early post-transplant growth: alternation of leaf shape in field environments. *Tobacco Science* **17**, 14–16.

Raper, C. D. and Johnson, W. H. (1971). Factors affecting the development of flue-cured tobacco grown in artificial environments. II. Residual effects of light duration, temperature and nutrition during growth on curing characteristics and leaf properties. *Tobacco Science* **16**, 31–35.

Raper, C. D. and Thomas, J. F. (1972). Temperature is early post-transplant growth: effect on shape of mature *Nicotiana tabacum* leaves. *Crop Science* **12**, 540–542.

Raper, C. D., Smith, W. T. and Downs, R. J. (1973). Growth responses of tobacco to light schedules in controlled-environment rooms. Paper No. 73–4526, *American Society for Agricultural Engineering, Annual Meeting*.

Rice, J. C., Gooden, D. T. and Price, E. L. (1970). Measured crop performance: Tobacco 1970. *NCSU Research Report* 36.

Vaartaja, O. (1954). Photoperiodic ectotypes of trees. *Canadian Journal of Botany* **32**, 392–399.

Wakeley, P. C. (1954). Planting the southern pines. Agricultural Monograph No. 18, Forest Survey, U.S.D.A.

Went, F. W. (1957). "Environmental Control of Plant Growth." *Chronica Botanica* **17**, Ronald Press, N.Y.

CHAPTER VII

Controlled Environments for Plant Research

A large number of prefabricated, commercially made plant growth chambers are in use and a general idea, usually somewhat biased, of the capabilities of such equipment can be obtained from manufacturers' brochures. There are also a large number of built-in, often special purpose, controlled-environment chambers. Many of these have been described in individual reports. Unfortunately a single publication describing the major controlled-environment facilities in Europe, or in Australia, New Zealand, or in North America is not available. However, Hudson (1957a, b) has described some seventy installations in England and in Japan a committee of scientists published "Environment-controlled growth rooms in Japan" in 1962 and in 1969 formed the Japanese Society of Environment Control in Biology. A supplement "Phytotrons and growth cabinets in Japan, 1972" was prepared for the UNESCO–NSF–SEPEL Phytotronics Symposium. In addition the journal of this society has contained many excellent articles relating to controlled environment.

Greenhouses

The original purpose of the greenhouse was to insure the survival of plants during the winter months. Plant cultivation in greenhouses was soon extended to all year and methods of cooling as well as heating began to receive attention. The development of plastics resulted in new greenhouse designs and enabled structures to be built when funds were not available for heavy duty glasshouses.

Improved technology and increased biological knowledge have resulted in more emphasis on climate control in greenhouses. Commercial greenhouse operators are using climate as a production tool for many crop types. Climate control in research greenhouses is rapidly becoming a necessity as the applied problems become more complex and the basic studies more sophisticated.

TEMPERATURE

Automatic ventilators and shades, fan and tube air distribution, and evaporatively cooled air contribute to better environment control. The best control, however, can only be obtained through mechanical refrigeration and it is only with mechanical refrigeration that full advantage can be made of atmosphere control. Unfortunately, very few mechanically air-conditioned greenhouses have been reported outside large phytotrons. The huge amounts of space required by the air-conditioning equipment, engineering most charitably described as poor, and the resulting high cost of the system have contributed to retard development of controlled-environment greenhouses. Research is needed to provide definitive information on greenhouse air-conditioning. Vast mechanical equipment space can cercertainly be reduced by careful systems design. Engineering can be improved but not by the current method of hiring those who have never progressed past air-conditioning of the local elementary schools. Greenhouse climate control is a true engineering problem and should be so treated.

Real information on greenhouse air-conditioning designs such as installed refrigeration capacity, air turn-over rates, inlet and discharge air temperature differences, etc. is not readily available. Discussions of how the designed system worked in practice do not seem to exist. Yet sizing of the refrigeration system seems to be a problem because of the radically different capacities used by various installations. Granted that solar insolation varies, summer climatic conditions and even glass type will alter cooling requirements, the differences between various designs seem so great that calculation methods are suspect. The SEPEL greenhouses have an installed cooling capacity of 87 700 kcal h^{-1} (29 tons where 3024 kcal h^{-1} = 1 ton) for a volume of 164 m^{-3}. This amounts to 544 kcal h^{-1} m^{-3} (0·18 tons m^{-3}) for a space design condition of 12°C. In practice the SEPEL greenhouse maintained 12°C when outside conditions exceeded the design values. Moreover, the test was made using a 9–10°C chilled water temperature instead of the 6·5°C intended by the specifications. In view of our operating experience it seems quite likely the SEPEL greenhouses are over-designed by a factor of 1·5 to 2. Kowalczewski (1963) commented that from calculations alone, without the prototype greenhouse, they would have installed 50% more cooling capacity at the Canberra phytotron. Greenhouse cooling requirements are calculated by the same rules as buildings for people and it seems some additional factors need to be considered.

Greenhouse cooling load calculations often seem unfathomable to biologists. For example, the calculated loads for the SEPEL greenhouses were:

Dry bulb	Wet bulb	r.h.	kcal h^{-1} m^{-3}	tons m^{-3}
24C	18C	60%	454	0·15
13C	9C	60%	544	0·18

Continuing the series, using the same calculation methods, to 7°C dry bulb 4°C wet bulb would result in a requirement of 564 kcal h^{-1} m^{-3} or 0·186 tons m^{-3}; an increase of about 6%. Yet the engineers insisted that the additional 5°C decrease would require 451 kcal h^{-1} m^{-3} (0·149 tons m^3); an increase of 83%. A direct expansion coil of 24·5 ton capacity sized for 5180 m min^{-1} was then placed in an air handler that would operate at 10 970 m min^{-1}. These facts simply serve to illustrate points about refrigeration calculations that average biologists like ourselves have difficulty in understanding. Such facts also provide an example where costs were much higher than necessary. After observing the operation of the SEPEL greenhouses it seems quite likely that 20–25°C temperatures could be maintained under full insolation and 38°C outside conditions with a cooling coil capacity of about 0·10 tons m^{-3}. The cost of heating and cooling over a range of 10–33°C need not exceed 40% of the greenhouse cost (Kowalczlewski, 1963).

LIGHT

Researchers using greenhouses are certainly aware of the disadvantages as well as the advantages of natural light. While the light quality may be more valid in terms of ecological thinking, the low winter irradiances are not representative of light levels during the normal growing season (Table I). In fact, many areas have so few sunny days during the winter months that acceptable plants cannot be obtained in greenhouses. Therefore the ecologist is faced with abnormal lighting conditions in the greenhouse even though the sun is the source. The physiologist is faced with poorly grown plants that have little validity for research.

TABLE I

Average daily total energy for various winter and summer months (cal cm^{-2} day^{-1}). Adapted from de Bilderling (1972)

Month	Location			
	Stockholm Sweden	St Maur France	Irkutsk Siberia	Washington U.S.A.
October	107	177	117	291
November	49	87	35	205
December	22	61	28	155
January	29	75	28	173
February	75	133	97	243
Winter mean	56·4	106·6	61·2	213·4
May	422	441	285	484
June	450	480	384	513
July	414	452	345	491
August	319	398	272	438
September	218	290	172	375
Summer mean	407·3	412·2	291·6	460·2

In northern latitudes a comparison of plant growth in greenhouses and controlled-environment chambers would favor the growth room most of the time. Examples for the month of September in North Carolina are shown in Table II.

Since light can be a limiting factor for plant growth in greenhouses, it follows that every effort should be made to insure maximum light transmittance. In the U.S.A. greenhouses are still being constructed with small glass panels whereas in Europe the Dutch-light glazing (71 × 142 cm) is preferred. Many phytotrons, including SEPEL, use large panels the order of 122 × 244 cm. Regardless of glass size dirt accumulation will drastically reduce the light transmission. Seeman (1951) reports that in industrial areas a 10 month accumulation of dirt and soot can reduce the illuminance by 56% and others (Van Koot and Djkhuizen, 1968) indicate reduction of light transmissions to 20–30% are not unusual. Loss of illumination in the greenhouse is usually a slow process and the user may be unaware of the amount of light lost until it reaches an obviously critical level for plant growth. Light transmission of many greenhouse structures is further reduced by the overhead location of pipes and electrical conduit arranged in the most convenient manner for the installer instead of following existing structural members. The structural members themselves cause large oscillations in light intensity. These oscillations can be smoothed considerably using water flow over the roof or by using cast glass.

TABLE II

Effectiveness of growth chambers and greenhouses during September for growth of several kinds of plants

	Fresh Weight (g)				
	Seneca Chief Corn	Brownie Scout Marigold	Manapal Tomato	Comanche Petunia	Happy Time White Petunia
Greenhouse[a]	50·0	15·0	23·7	14·2	11·7
CER, 9 h[a]	54·1	21·3	38·1	21·0	15·8
CER, 15 h	95·4	36·4	54·8	29·4	20·7
Days from seed	25	32	32	35	35

Coker 319 and 254, 19 days from seed

	Leaf length (mm)		Leaf width (mm)		Fresh weight (mg)	
	C-319	C-254	C-319	C-254	C-319	C-254
Greenhouse[a]	11·8	15·5	9·3	12·2	52	78
CER, 9 h[a]	23·4	26·0	19·0	21·1	149	174
CER, 15 h	30·8	36·6	25·0	26·4	399	493

Cherry Belle Radish, 18 days from seed

	Fresh weight (g)		Root diameter (cm)	Root dry weight (mg)
	Total	Root		
Greenhouse[a]	4·0	0·7	0·77	32
CER, 9 h[a]	6·5	1·9	1·36	100
CER, 15 h	6·9	3·1	1·74	203

Marketer Cucumber

	Largest leaf			Stem length (cm)		Weight (g)		Chloro-phyll (mg g^{-1})
	Length (cm)	Width (cm)	Nodes	Ground to cotyl	Cotyl to tip	Fresh	Dry	
Greenhouse[a]	12·2	14·1	7·5	9·4	35·2	30·2	2·51	0·161
CER, 9 h[a]	13·2	15·5	8·0	7·8	41·2	36·0	2·90	0·185
CER, 15 h	12·8	16·1	11·0	2·4	48·2	47·1	4·19	0·178

[a] 3 h dark period interruption

Use of artificial light to supplement natural lighting in the greenhouse suffers from two major disadvantages: (1) the large electrical load required to obtain enough illumination to make a significant contribution and (2) the lighting system can create considerable shade and further reduce the available natural light. Generally greenhouse supplementary lighting has not met with tremendous success, at least in part because the additional irradiance has been insufficient. In many areas, of course, the natural winter light is so poor that any additional illumination improves plant growth and with some commercial crops such as roses the result appears to be economically profitable. Pinchbeck *et al.* (1971) reported that 100 hlx of supplementary light increased rose production 73% with a return on investment in excess of 30%.

The phytotron in Stockholm probably has the best greenhouse lighting system yet devised. There a fluorescent light source is mounted on "telpher" cars which are moved into and out of the greenhouse automatically (Wettstein, 1967). Another system reported by Smith *et al.* (1973) uses a metal halide supplementary source that moves back and forth along the greenhouse bench at a rate calculated to supply the photosynthetic needs during the light-deficient winter months.

Photoperiod control in greenhouses is usually accomplished by black cloth shading to obtain days shorter than normal and by extending the natural day or interrupting the dark period with low intensity artificial light for long day effects. In some phytotron applications the entire greenhouse may be automatically darkened (Wettstein, 1967) or small individual glass-walled cabinets are provided with light-tight covers as at Canberra (Read *et al.*, 1963). In other cases plants are moved from the glasshouses to controlled-environment rooms each day as at Cal Tech (Went, 1957) or to photoperiod rooms which provide only a low level illuminance as at Beltsville (Borthwick, 1946). None of the methods is completely satisfactory because the automatically closing systems need considerable maintenance and the photoperiod room method requires the plants to be moved twice daily. Photoperiod rooms, however, have proven over the years to be the least costly method initially, with the least maintenance and downtime for repairs.

Plant Growth Chambers

The plant growth chamber might be described as simply an artificially lighted, insulated box in which temperature- and humidity-controlled air is circulated at about 30 m min^{-1}. Satisfactory design of such controlled-environment facilities is not particularly difficult, yet poor to terrible performance histories are commonplace. Part of the problem stems from inadequate specifications written by biologists with little engineering experience or by engineers with even less understanding of biology. Part of the problem also stems from insufficient to a total lack of maintenance.

The biologist usually buys or builds plant growth chambers on the basis of the performance necessary for his program of use plus a little extra for what he might need some day. Attempting to include possible future requirements simply runs up the cost and rarely succeeds. Efforts have been made to assist the biologist in preparing performance requirements (AIBS, 1971; Acock, 1972).

Unfortunately, after the chambers are put into operation, the initial performance, even if it meets specifications, usually deteriorates. Malfunctions that destroy experimental material and effort then increase in frequency, severity and duration. These things do not have to happen, but quite often the individual chamber owner lacks the time and assistance to set up and run a comprehensive maintenance program, especially when that job is complicated by designs that never seriously considered maintenance methods. In cases where Maintenance and Operations or Physical Plant personnel are used to care for the equipment results have rarely been satisfactory. These much maligned and usually overextended organizations, with very few exceptions, cannot operate on the emergency repair basis needed by controlled-environment facilities. Nor do they have the manpower to provide a high-level maintenance program.

Characteristically few individually owned growth chambers are provided with any maintenance program other than sporadic lamp changes. Instead the system is cared for on the "fix it when it stops" procedure which often repeatedly repairs the same flaw instead of redesigning it to provide a permanent solution. As an example, chambers have been constructed (not at our institutions, of course) where lamp ballast life was reduced to 25% of normal because of overheating due to improper mounting. Local mechanics "solved" the problem by installing drip pans to prevent the ballast potting compound from dripping on walls, floor and heads of personnel. Several years later the user was still plagued by lamp outages caused by ballast failures; yet the cost of a single year's ballast replacement would have paid for remounting the ballasts to provide a more normal life.

Irrespective of maintenance problems the individually owned plant growth chamber is frequently not used efficiently. An inspection of such chambers at almost any University or large research station nearly always reveals empty controlled-environment rooms, while other researchers at the same location are trying to get funds for similar equipment.

In spite of the problems and limitations of individually owned controlled-environment rooms they have contributed to much important research. As the designers pay more attention to maintenance and to long-term performance, contributions from such facilities will increase. Nevertheless, they will remain inefficient and cannot overcome the disadvantage of the limited environmental parameters.

Phytotrons

OPERATION

The more ideal method of operating controlled environment facilities is to accumulate a large number of plant growth chambers in one location under a management team. However, this arrangement in itself does not constitute a phytotron. Hudson (1957) defined a phytotron as "a complex installation which includes a series of growth rooms, temperature-controlled rooms and glasshouses where plants can be moved freely from one environment to another". Berrie (1957) improved the definition by pointing out that a "phytotron is as much a system of management as a collection of growth rooms since the flexibility of the phytotron is achieved in large measure by the management". Because the philosophy of operation is involved, any discrete definition of a phytotron would probably fail to obtain universal acceptance. However, a general definition might be that a phytotron is a laboratory designed primarily for studying the response of plants to their environment and is organized so that many combinations of environmental factors can be studied simultaneously. The facilities obviously should provide uniform environmental conditions, adjustable over a wide range.

The advantages of a phytotron over individually owned plant growth chambers are obvious. The investigator is able to study a greater number of environmental variables simultaneously. The larger facility can justify a full-time maintenance program conducted by expert refrigeration and electronic technicians. Biological research requirements are rarely static and phytotron installations can supply the engineering skills to meet changing needs. Phytotrons can be operated more efficiently because the research space is available to many rather than to a single owner. As a result the space is nearly always occupied. The facilities in the SEPEL are constantly in use, yet the turnover and input is balanced so that the space requests can be met with very little delay. Admittedly the balance is at times precarious and additional facilities are needed to alleviate it and provide a small excess that can be shut down for maintenance and repairs.

Criticisms of phytotrons often center around the term "inflexibility". The SEPEL attempted to overcome this objection by using greenhouses and several sizes of dark rooms, germinators and plant growth chambers. Plants can be moved twice daily between greenhouses and dark rooms or growth chambers, or between the large and medium-sized growth rooms. Smaller reach-in chambers are used for small plants, seedlings, insects, etc. where movement is undesirable or unnecessary. Still smaller mini-chambers are currently being designed especially to accommodate studies with insects, algae, hydra, snails, etc.

The SEPEL phytotrons are not static. The rooms can be dismantled, if necessary. Several of the B-chambers could be combined into one large room. The lighting systems can be completely removed and if desirable replaced with a different, more advanced system, when and if it becomes available. Additional space was included in which future, perhaps very special purpose, controlled-environment chambers can be installed (Table III). For example, the Phytocyclon of Wolf (1969) would be a very useful

TABLE III

Characteristics of the Southeastern Plant Environment Laboratories (SEPEL)

	NCSU unit	Duke unit
Building area	3922m²	3192m²
Building volume	18589m³	15175m³
Electrical supply	480/277 V, 3 ph	Y connected
Present connected load	2240 KVA	1810 KVA
Possible connected load	3000 KVA	2000 KVA
Refrigeration		
Chilled water	372 MBH*	298 MBH
Building	228 MBH	182 MBH
Greenhouses	144 MBH	116 MBH
Ethylene glycol		
Effective capacity	414 MBH	276 MBH
Expansion capacity by dilution	480 MBH	320 MBH
Controlled environment rooms		
Greenhouses	3	6
Cooling	36 MBH	36 MBH
Lights	Photoperiod control	both units
A–chambers 2·44 × 3·66 m	22	8 light 2 dark
Cooling	9 MBH	9 MBH
Fluorescent lamps	18060 W	18060 W
Incandescent lamps	4800 W	4800 W
B-chambers 1·22 × 2·44 m	10	10
Cooling	4 MBH	4 MBH
Fluorescent lamps	6020 W	6020 W
Incandescent lamps	2400 W	2400 W
C-chambers 0·91 × 1·22 m	20	20
Cooling	0·6 MBH	0·6 MBH
Fluorescent lamps	1600 W	1600 W

	NCSU unit	Duke unit
Building area	3922m²	3192m²
Building volume	18589m³	17175m³
Incandescent lamps	600 W	600 W
Seed germinators	12	
	No load data	
Incubators	2	
Cooling	0·6 MBH	
Dew chambers	2	
Cooling	0·6 MBH	
Air pollution treatment rooms		
Six photoperiod rooms		

Expansion capacity in the future

NCSU	Duke
Two greenhouses	Three 2·44 × 3·66 growth rooms no height limit
Three 2·44 × 3·66 m growth rooms up to 6 m high	
Six 1·22 × 2·44 m growth rooms	Four, 1·22 × 2·44 m growth rooms
Eight 0·91 × 1·22 m growth rooms	Five 0·91 × 1.22 m growth rooms
Four 1·88 × 2·44 m dark rooms	
8–10 special chambers for insects	
6 photoperiod rooms	

Note: MBH is thousands of British Thermal Units per hour 1 MBH = 252 cal; kg^{-1} = 8·33 × 10^{-2} tons of refrigeration. (U. S. comm.)

adjunct to our facilities. This device is equipped with separate root and shoot environments. Simultaneous and continuous measurements can be made of whole plant assimilation, respiration and transpiration as functions of various environmental parameters such as temperature, irradiance or CO_2 concentration.

Most of the phytotrons have been described (see References) and some seem to be more flexible than others. Some have special equipment, e.g. the spectral illuminator at Gif (Chouard *et al.*, 1972), the monochromator at the Institute of Physical and Chemical Research, Saitama Japan (Kawarada, 1971), and the biological spectrograph at the Biotron Institute, Fukouka (Matsui *et al.*, 1971).

RESEARCH PATTERNS

Organization of research in phytotrons is based on the objectives of the facility at the time it was constructed. The French phytotron at Gif-sur-Yvette is an example of phytotron-organized research. At Gif a permanent phytotron staff forms a collection of teams, each with a definite region of activity. One team is studying the factors influencing flowering while another is investigating the biochemical activities that accompany various morphogenic changes brought about by the environment. An ecological group is attempting to understand competition within plant communities and another is studying how environment modifies the mineral nutrition complex. Visiting scientists, whether from France or abroad, join with the permanent research team working in their area of interest.

In SEPEL the least internal organization is used. Here the individual investigators submit research programs that may consist of only a single experiment in the phytotron. Usually the program is a complex of studies including field and greenhouse research although that fact may not be readily apparent from the phytotron phase. Different investigators pursue programs which, while not correlated at the phytotron level, are part of departmental organized research. Visiting scientists in this type of organization can conduct investigations individually in cooperation with departmental research teams (Chouard *et al.*, 1972).

All phytotrons are or should be engaged in the study of phytotronics, the technology involved in the use and design of phytotrons. For example, if you had a light source that produced an illuminance of 10–1200 hlx or one that provided a total irradiance of $1 \cdot 5$ cal cm^{-2} min^{-1} would plants make any real use of the energy over what is now available? The answer would be "no", unless (1) the spectral distribution were acceptable and (2) the radiant heat were reduced to tolerable levels. There is little evidence that plants, even when all precautions are taken, would use solar level energies efficiently enough to justify the cost. However, the lack of evidence may be due to lack of investigation and this is one of many areas of phytotronics that might receive attention.

DESIGN AND ENGINEERING

Phytotronics is intimately concerned with the design of phytotrons which include, of course, single plant growth chambers. The reason for this concern was noted by Went in the preface to *Phytotronique* I (Chouard and de Bilderling, 1969) when he stated:

> "my experience with Phytotrons has shown that one of the most important conditions in their establishment is the collaboration with a fully competent engineer. Most air conditioning engineers have had no experience with the controlling of the greenhouse (or plant growth chamber) environment and consequently grave errors have been made in the construction of a number of them".

Architects are selected for phytotron design by whatever method prevalent at time: rotation among local firms, reputation for church or office building design or other vague and irrelevent criteria. The architect then selects an engineer by some method not immediately clear to anyone outside the profession. Apparently, if an engineering firm has been involved in any kind of controlled environment construction they become an instant authority on the subject and are therefore imminently qualified to do the engineering of a phytotron. One thing the architect definitely does not do is ask the owner or user how well the engineer did the previous job. Apparently, this is because owners are usually disgruntled, impossible to please and hardly qualified to render an opinion. The result is as Went pointed out—too often a collection of grave errors that are largely unnecessary. Even in cases like the phytotron at Canberra where a rather large group of truly competent engineers were gathered together, errors were avoided only by construction of prototypes. Unfortunately the experience gained was then largely lost as the group dispersed to other projects.

The experience of phytotron design rests almost entirely with those actively working full-time in such capacities as directors, engineers or research associates. These are the people who know the advantages and disadvantages of the various systems used to make up a phytotron, where errors have been made and where new ones are most likely to occur.

Many of the errors found in existing phytotrons are difficult to explain: e.g. an air handler sized to deliver $1020 \text{ m}^3 \text{ min}^{-1}$ that is attached to a duct sized to handle $480 \text{ m}^3 \text{ min}^{-1}$; an electrical supply to a group of growth chambers that includes a diversity factor for the lighting systems as if they were a group of wall outlets; or space air conditioning that fails to include the heat emitted from the plant growth chamber lighting systems. Yet a phytotron free from design errors is quite possible. An ideal phytotron for general use could be described in sufficient detail that it could be built without major flaws. Special-purpose phytotrons for one kind of plant or special areas of research would differ only in minor ways not in the basic features. For example, all mechanical and electrical equipment must be readily accessible and removable without disrupting key components: an obvious design parameter that the engineer would surely consider automatically. Unfortunately, examples are available where the entire refrigeration system must be shut down to repair a leaky valve welded into the line without benefit of unions or flanges. Failure to consider such small details severely hampers continuity of operation and the ideal phytotron would prevent such errors.

The units that make up a phytotron—the plant growth chambers—can also be described in ideal detail. Nearly all the design errors that currently occur in controlled-environment rooms could have been avoided. In many growth rooms it is necessary to disrupt the research in order to change lamps or fans, yet where such disruptions are simply not acceptable workable systems have been developed: lamp access systems have been designed (Doorenbos, 1964; Bretschneider-Herrmann, 1969) and the equipment at SEPEL was modified to allow rapid replacement of fans without any system down time.

Much of the failure to maintain performance can be traced to the controls, which is hardly surprising when one considers that habitually this key system amounts to less than 5% of the cost of the controlled-environment room. Using better quality control components, perhaps worth 10–15% of the total room value, would reduce malfunctions by at least a factor of 10.

The Southeastern Plant Environment Laboratories are examples of phytotrons planned, constructed and put into full operation in a minimum time. The two buildings represent somewhat different problems since one is multi-story and the other essentially a single floor work area plus a mechanical area. Quite unbelievable design errors were found and corrected prior to construction simply by having the phytotron directors check the plans and calculations presented in the various sets of preliminary and working drawings. The first set of drawings and specifications would, in our opinion, have resulted in an inoperable facility costing considerably more than our budget. The revised plans with most of the errors removed resulted in an operable system close enough to the budgeted funds to allow construction. The large amount of time spent in planning and checking the calculations and specifications paid off in terms of reduced running-in time and overall successful operation.

A properly operated phytotron is never completed. Changes are constantly being made to improve and enlarge upon the environment control capabilities. As research trends and priority for information shift to new areas the phytotron must be prepared to accomodate the new demands. Therefore funds must be made available for new equipment, modifications and improvements if a phytotron is going to have a productive life of more than 10 years.

PHYTOTRON *vs* INDIVIDUAL OR SMALL GROUPS OF CHAMBERS

Nitsch (1972) made a critical appraisal of achievements of phytotrons, pointing out the high costs of operation and the lack of use of the Biotron in Wisconsin in 1969. However, a visit made by us to several phytotrons in Europe in 1967 left the impression that most units were being used extensively. Furthermore, our experience with the use of the two units of the Southeastern Plant Environment Laboratories (SEPEL), located at Duke University and North Carolina State University, indicates that not only is the use heavy but it is rising rapidly. In 1969 a total of 75 researchers used the facilities and by 1972 the number increased to 105. This increase also corresponded with an increase in complexity of the experiments and the efficiency of the use of the space as the researchers learned more about the capabilities of the phytotrons.

In contrast to the isolated unit an analysis of the situation at SEPEL showed that many advantages accrued to the scientist who uses the phytotron.

(1) The scientist can start his experiment almost immediately. He does not have to wait until a chamber is purchased and installed. (2) Availability of a large number of controlled environments simultaneously. This saves time in conducting a multi-environment experiment. (3) The scientist does not spend valuable and expensive time operating and maintaining the equipment; this is done by phytotron personnel. (4) Excellent reliability records due to preventive maintenance and experienced personnel, show that an experiment in all probability will be completed without loss of plants due to mechanical failure. (5) The scientist is not burdened with an expensive piece of equipment that he feels obliged to use even though his critical experiment has been completed.

In summary, the individual scientist using the phytotron saves considerable time and money as compared with developing and operating his own system of a few chambers. This has been found to be true even if the cost of commuting a considerable distance to set up, make monthly observations and to harvest the plants is included.

References

Acock, B. (1972). A prototype airtight, daylit cabinet and the rationale of its specifications. In "Crop Processes in Controlled Environments" (A. R. Rees, K. E. Cockshull, D. W. Hand and R. G. Hurd, eds), pp. 91–107. Academic Press, London and New York.

AIBS Bioinstrumentation Advisory Council (1971). Controlled environment enclosure guidelines. BIAC Information Module M21.

Berrie, A. M. M. (1957). Phytotrons. In "Control of Plant Environment" (J. P. Hudson, ed.). Butterworths, London.

Borthwick, H. A. (1946). Photoperiodic response as a factor in choice of plants for testing soil deficiencies. Soil Science 62, 99–107.

Bretschneider-Herrmann, B. (1969). The phytotron in Rauisch-Holzhausen: technical details and experiences. "Phytotronique I", pp. 24–26, Centre Nationale de la Recherche Scientifique, Paris.

de Bilderling, N. (1972). Phytotrons et environment dans les espaces climatises. "Phytotronique II). (P. Chouard and N. de Bilderling, eds), pp. 15–55. Gauthier-Villars, Paris.

Chouard, P. and de Bilderling, N. (eds) (1969). "Phytotronique I." Centre National de la Recherche Scientifique, Paris.

Chouard, P., Jacques, R. and de Bilderling, N. (1972). Phytotrons and phytotronics. *Endeavour* **31**, 41–45.

Committee for Environment-controlled Growth Rooms in Japan (1962). "Environment-controlled growth rooms in Japan."

Doorenbos, J. (1964). The phytotron of the Laboratory of Horticulture, State Agricultural College, Wageningen. *Overdruk. Med. Dir, Tuinb.* **27**, 432–437.

Hudson, J. P. (1957a) Control of plant environment for experimental work. *University of Nottingham Department of Horticulture Miscellaneous Publications* **8**.

Hudson, J. P. (1957b). "Control of Plant Environment." Butterworths, London.

Japanese Society of Environment Control in Biology (1972). "Phytotrons and growth cabinets in Japan."

Kawarada, A. (1972). Phytotron, Institute of Physical and Chemical Research. *In* "Phytotrons and Growth Cabinets in Japan", pp. 87–93.

Konishi, M. (1972). Phytotrons in Japan and the Japanese Society of Environment Control in Biology and its activities, including the plan of the National Biotron Center. *Environmental Control in Biology* **10**, 1–10.

Kowalczewski, J. J. (1963). Phytotrons. *In* "Engineering Aspects of Environment Control for Plant Growth", pp. 122–131. C.S.I.R.O., Australia.

Matsui, T., Aiga, I., Eguchi, H. and Asakawa, F. (1971). Biological studies on light quality in environment control II. Biological spectrograph: with special reference to operational characteristics of the instrument. *Environmental Control in Biology* **9**, 111–119.

Nitsch, J. P. (1972). Phytotrons: past achievements and future needs. *In* "Crop Processes in Controlled Environments". A. R. Rees, K. E. Cockshull, D. W. Hand and R. G. Hurd, eds), pp. 33–35, Academic Press, London and New York.

Pinchbeck, W., Johnson, F. K., Stiles, D. N. and Noessen, S. J. (1971). Increased production of Forever Yours Roses with supplemental lighting. General Electric Technical Information Series 71–0L–001.

Read, W. R., Cunliffe, D. W., Chapman, H. L. and Kowalczewski, J. J. (1963). Naturally lit plant growth cabinets. *In* "Engineering Aspects of Environment Control for Plant Growth", pp. 102–122. C.S.I.R.O., Australia.

Seeman, J. (1951). Untersuchungen zur Verschutzung von Glasflachen. *Tatigkeitsber. d. Gartn. Versuchsenstatt Friesdorf.* **23**.

Smith, W. T., Downs, R. J. and Jividen, G. M. (1973). Economical HID source for greenhouse light supplement. Abstract, ASHS Annual Meeting.

Van Koot, I. Y., and Dijkhuizen, T. (1968). Light transmission of dirty glass and cleaning methods. *Acta Horticulturae* **6**, 97–107.

Wettstein, D. von. (1967). The Phytotron in Stockholm. *Studia Forestalia Suecica*, **44**, 1–23.

Wolf, F. (1969). New climatic measuring chambers for plant physiological research: technical description. *Phytotronique* **I**, 12–16.

Went, F. W. (1957). "Environmental Control of Plant Growth." Chronica Botanica 17, Ronald Press, New York.

Results of Research Conducted
in Controlled Environments

The usefulness of research results, either for further research or practical application, is the real criterion for evaluating research. To make the results from controlled environment research available in a form that is useful requires that the conditions under which the research was conducted be described as fully as possible. The American Society of Horticultural Science (ASHS) has published a set of guidelines on reporting such results (Committee on Growth Chamber Environments, 1972).

It was proposed that the information be presented in paragraph form similar to the following model:

Studies were conducted in a (make and model) chamber with a (no, type of material) barrier having (   )% input wattage of (brand) (type) lamps and (   )% (brand) (type) lamps. The average luminous flux density (often erroneously called light intensity) during the study at the top of the plants was (   ) (klx) at the beginning and (   ) (klx) at the termination as measured with a (make and model) meter with a cosine corrected filter. It would be better of course to give average photon flux density of PAR. Light duration was (   ) hours with light–dark (abrupt or gradual) change. The air temperature at the top of the plants was (degree) $C \pm$ (   ) during the light period and (degree) $C \pm$ (   ) during the dark as measured with a (thermocouple, thermistor, etc.). Average temperature at the center of the (number) containers was (   ) $C \pm$ as measured with a (mercury thermometer, thermistor, etc.). Relative humidity was maintained at (   )% $\pm$ (   )% during the light period and (   )% $\pm$ (   )% during the dark. The airflow was (up, down or across) the chamber and at a rate of (   ) meters per minute at the top of the plant canopy as measured with a hot-wire anemometer. Fresh air at the rate of (   )% per hour of the total volume. [CO_2 level was monitored and controlled with an infrared analyzer and maintained above 300 ppm.]

Plants were grown in (media) in (size) (type: plastic, clay) containers. The plants were (transplanted or grown from seed) in (conditions) and to (size) before the start of the experiment. Plants were irrigated (how

141

frequently) with (distilled, deionized, tap) water and/or (strength) nutrient solution. (Give composition or reference to published composition.)

Following these guidelines may not always be practical because many investigators lack the instrumentation. Moreover, no standards are available that will insure each investigator makes measurements in the same way. Nevertheless, the ASHS guidelines are a beginning and if followed as closely as possible, will be a major step in advancing uniform reporting and thus biological results will be more accurately evaluated and interpreted by the reader.

Reference

Committee on Growth Chamber Environments, American Society of Horticultural Sciences (1972). Guidelines for reporting studies conducted in controlled environment chambers. *HortScience* 7(3), 239.

Index

143